Training Requirements and Training Delivery in the Total Army School System

John D. Winkler
John F. Schank
Michael G. Mattock
Rodger A. Madison
L. Diane Green
James C. Crowley
Laurie L. McDonald
Paul S. Steinberg

Prepared for the
United States Army

Arroyo Center

RAND

Approved for public release; distribution unlimited

For more information on the RAND Arroyo Center, contact the Director of Operations, (310) 393-0411, extension 6500, or visit the Arroyo Center's Web site at http://www.rand.org/organization/ard/

PREFACE

In recent years, the U.S. Army has launched a series of initiatives to streamline and consolidate its extensive system of schools, covering training institutions that serve both active and reserve forces. Prominent among these initiatives is a prototype regional school system the Army established in the southeastern region of the United States during fiscal years 1994 and 1995, which fundamentally changed the organization and management of Reserve Component Training Institutions run by the Army National Guard (ARNG) and the U.S. Army Reserve (USAR). The objectives of this initiative were to achieve economies and ensure the quality of training, while laying the foundation for a "Total Army School System" (TASS) that would be more efficient and integrated across the Active Component (AC) and the Army's two Reserve Components (RC).

As this reorganization got under way, the RAND Arroyo Center was asked to provide an objective assessment of the performance and efficiency of the Army's system of schools, including the regional prototype. This document presents final results for one of the major areas in the assessment, which examined school system ability to meet training requirements. A companion report, MR-844-A, *Resources, Costs, and Efficiency of Training in the Total Army School System,* examines resource use and efficiency of training, while MR-955-A, *Performance and Efficiency of the Total Army School System,* provides an overall summary of the Arroyo Center's final results and recommendations.

This report documents the key findings and recommendations from these analyses for historical purposes. Some of the recommendations were adopted and are so noted in the report. Other recommendations, also noted here, have not yet been adopted. Hence these results can still be used to guide the further development of the TASS.

The research was sponsored by the Deputy Commanding General, U.S. Army Training and Doctrine Command, and the Commanding General, U.S. Army Combined Arms Center, and was conducted in the Arroyo Center's Manpower and Training Program. The Arroyo Center is a federally funded research and development center sponsored by the United States Army. This report should be of interest to policymakers responsible for defense manpower and training and for Active and Reserve Component issues.

CONTENTS

FIGURES

TABLES

SUMMARY

INTRODUCTION

This report analyzes training requirements and school delivery of training in the Total Army School System, focusing on the system's ability to meet its training requirements in Reserve Component Training Institutions. Two types of training are the subject of this research. The first is the education of Reserve Component noncommissioned officers (NCOs). The second is reclassification training of soldiers who previously held one military occupational specialty (MOS) but who now need to be trained and qualified to hold a different MOS.[1] The report examines the execution year of Region C, where the "prototype" Total Army School System was established in the southeastern United States (FY95)—compared to the implementation year (FY94), as well as examining the prototype in relation to the overall system. The document also analyzes the impact of personnel policy changes that could enhance the training system's flexibility and effectiveness in meeting training requirements (i.e., by lowering turbulence to reduce demands on the system). This report is part of a larger effort by RAND Arroyo Center to analyze the performance and efficiency of the RC school system.

---

[1]RC soldiers can require reclassification training if they change from one MOS to another while serving in the RC, or if they join the RC after having served in the AC in a different MOS.

TRAINING REQUIREMENTS AND CAPACITY UTILIZATION FOR NCO EDUCATION

This part of the analysis deals with the system's ability to manage its requirements and deliver training in its Noncommissioned Officer Education System (NCOES) across the two fiscal years, focusing on soldiers requiring training in the Primary Leadership Development Course (PLDC, required for promotion from grade E-4 to E-5); the Basic Noncommissioned Officer Course (BNCOC, required for promotion from grades E-5 to E-6); and the Advanced Noncommissioned Officer Course (ANCOC, required for promotion from grades E-6 to E-7). We find that across the fiscal years:

- **Training requirements are large but decreasing.** Our data from FY95 show approximately 71,000 soldiers in grades E-4 through E-7 in the ARNG and USAR needing to complete PLDC, BNCOC, or ANCOC for recent or impending promotion. Although large, this number is smaller than in FY94, when 84,500 soldiers needed the NCOES courses. The requirement fell for two reasons: (1) the Army continued to enforce its "select-train-promote" policy for NCOES, which emphasizes that only NCOs selected for promotion should attend NCOES classes; and (2) the "backlog" of untrained NCOs fell because many of these NCOs received the required training for current or impending grade or left the force.[2] If these trends continue, the NCOES training requirement should continue to fall in the direction of its "steady state," in which the only NCOs requiring NCOES are those promoted (about 8 percent, or 20,000 NCOs in grades E-4 through E-6 per year, at current force levels).
- **Training capacity is better able to meet demand.** There continues to be an imbalance between the current training requirement (71,000 NCOs) and the school system's ability to meet this requirement (44,500 school quotas, or 63 percent of the requirement), but this is better than in FY94, where the system had enough seats programmed to meet 55 percent of the requirement (46,500 quotas for the 84,500 NCOs needing training). Although there were more quotas in FY94, the requirement was

[2]Force structure decreased during this period but at a proportionately smaller rate.

proportionately higher. As training requirements continue to fall, continued and more extensive consolidation should be possible within the part of the Army school system that provides NCO training.

- **Inefficiencies in using quotas grew worse.** Quota utilization actually grew worse in FY95 as compared to FY94, driven by a dramatic increase (11 percentage points) in unfilled seats. Altogether, the quota utilization in PLDC and NCOES Phase 2 declined by 10 percentage points and 5 percentage points, respectively. Aside from the growing problem of unused quotas, further inefficiencies remained because some of those who attended the courses appeared not to need them. About twice as many soldiers in grades E-4 through E-6 attended the NCOES course required for the next-higher grade than were promoted in FY95. Similarly, about half the E-4s attending PLDC, E-5s attending BNCOC Phase 2, and E-6s attending ANCOC Phase 2 in FY94 still remained in these grades in FY95.
- **Production of graduates decreased slightly.** Although the RC school system offered NCOES quotas for a larger portion of the training requirement in FY95 than in FY94 (63 percent versus 55 percent), it did worse in using quota allocations (63 percent versus 71 percent) and in producing graduates (84 percent versus 89 percent). The end result was that the overall system did marginally less well at producing graduates in relation to its training requirement across fiscal years (33 percent versus 35 percent). Plus, given that some students who received training appeared not to need it, the portion of the "true" NCOES training requirement that was met was even smaller.
- **Region C is comparable to the rest of the nation.** Despite these declines in school capacity utilization and production for Army NCO education, school system performance in the prototype was about the same as the rest of the nation on most of these measures. For example, in terms of training requirements, it did as well as other regions in implementing "select-train-promote" and better in reducing the "backlog" of nonqualified NCOs; in terms of quota utilization, it did about the same.

RECLASSIFICATION TRAINING REQUIREMENTS AND SCHOOL DELIVERY

This part of the analysis addresses duty MOS qualification (DMOSQ) training in the RC school system across the fiscal years. We find that:

- **Requirements decreased and capacity increased.** Although the training requirement is still large in FY95—75,543 at the start of the year—it is smaller than the initial FY94 requirement of 87,985. In addition, the capacity to meet this requirement also improved, with 36,631 initial quotas (48 percent of the initial requirement) in FY95 versus 31,619 quotas (36 percent of the initial requirement) in FY94.
- **Problems remained with utilization of school capacity.** Quota use in FY95 remained about the same compared to FY94 (69 percent versus 67 percent used) but is still low in absolute terms, and problems remained with unfilled (17 versus 18 percent) and cancelled classes (14 versus 15 percent).
- **Production of graduates improved.** Given reduced requirements, increased capacity, and equivalent quota utilization in FY95, the RC school system produced a larger number of graduates—23,758 versus 19,933. However, this output is still small in relation to overall training needs, amounting to 31 percent of the 75,543 requirement (versus 23 percent in FY94).
- **Region C did better than other regions in the nation.** Compared with other regions in the nation, Region C fared quite well in terms of DMOSQ training. It filled 81 percent of its quotas (versus 68 percent in other regions), had a substantially smaller percentage of unfilled training seats (3 percent versus 19 percent), and had a better percentage of graduates in relation to quota allocations (71 percent versus 64 percent).
- **DMOSQ requirements actually rose by the start of FY96.** Although DMOSQ requirements fell from the start of FY94 to the start of FY95, they actually got worse again by the start of FY96. In spite of increased capacity and output, the number of soldiers showing a need for reclassification training rose from 75,543 at the start of FY95 to 82,166 at the start of FY96. The end result was that the DMOSQ rate fell from about 78 to 75 percent of assigned

personnel. The reason for the rising requirements and drop-off in DMOSQ rate derives from turbulence within the RC, which makes it difficult for the training system to make headway against the training requirement.

EFFECT OF REDUCING TURBULENCE ON THE DMOSQ TRAINING REQUIREMENT

Given these results, which reflect long-standing research results that show job turbulence to be an endemic and persistent problem in the U.S. Army Reserve Components,[3] we looked further into how reducing turbulence might affect the rate of MOS qualification and the DMOSQ training requirement.

Turbulence Is Personnel, Not Force Structure, Driven

Determining how to address the problem requires understanding the source of the turbulence we saw: Is it a product of force structure changes taking place during the time studied (1994–1995), or is it simply an endemic problem of "personnel churn"? We conducted a series of analyses examining how the personnel flows differ for units being activated, inactivated, or converted in some way versus those units that are "stable." We determined that personnel movements resulting from individuals' decisions to change jobs or leave the force, much more so than force structure changes, drive the turbulence we saw. This means the problem can be addressed by making changes to personnel policies that reduce personnel turbulence (e.g., "stay-in-place" incentives) and, thus, have an impact on attrition and voluntary job changes that generate a reclassification training requirement.

---

[3]Richard J. Buddin and David W. Grissmer, *Skill Qualification and Turbulence in the Army National Guard and Army Reserve,* Santa Monica, CA: RAND, MR-289-RA, 1994; Ronald E. Sortor, Thomas F. Lippiatt, J. Michael Polich, and James C. Crowley, *Training Readiness in the Army Reserve Components,* Santa Monica, CA: RAND, MR-474-A, 1994.

Reducing Turbulence and Attrition Significantly Increases DMOSQ Rate

We built an inventory projection model to estimate the future steady-state DMOSQ rate and used that rate as the baseline to estimate how reducing personnel turbulence would affect the demand for training. This model replicated analyses conducted in earlier RAND research (using a different model) that estimated the effects of reducing personnel turnover on RC readiness (Orvis et al., 1996). Like this earlier research, our work focused on two particular cases of turbulence reduction: (1) reduce job changes of duty-qualified soldiers by 50 percent; and (2) decrease attrition by 25 percent (in addition to reducing job changes of duty-qualified soldiers by 50 percent). For the first case, the DMOSQ rate increased from the estimated steady-state rate of 74.8 percent to 78.8 percent; in the second case, the rate increased further to 80.4 percent. At the same time, requirements for reclassification training decreased significantly, dropping by 19,000 soldiers in the first case (from about 81,000 to 62,000) and by about 20,500 soldiers in the second case (from 81,000 to 60,500, which is much closer to current training capacity).

CONCLUSIONS AND RECOMMENDATIONS

The results indicate that progress has been made within the prototype and within the broader system in managing DMOSQ training requirements and delivering the training needed to meet them. Capacity utilization and production of graduates seem to be improving throughout the system and particularly in Region C. At the same time, the scope of problems that remain warrants continuing vigilance and effort to solve them. These problems—common to both areas examined in this report—center around improving quota utilization (particularly for NCOES) and reducing the number of personnel needing training for both NCO education and reclassification training.

In terms of quota utilization, the system is wasting about a third of the training seats allotted to deliver training. These quotas are being lost for several reasons: (1) despite improvements, unit personnel responsible for making and monitoring reservations for training seats are not yet fully proficient in using the Army's reservation system (ATRRS); (2) key resources needed to conduct a course, including qualified instructors and the equipment and facilities, are absent; and (3) some soldiers who make reservations do not show up and additional available seats remain unfilled.

As for the first reason, responsible Army agencies (including the U.S. Army's Office of the Deputy Chief of Staff for Personnel, the U.S. Army Reserve Command, and the Army National Guard) are providing additional training and assistance in using ATRRS, and these efforts should be maintained. In addressing the second reason, organizations within the TASS responsible for coordinating training and certifying instructors, such as TRADOC's Regional Coordinating Elements (RCEs), can help ensure that key resources are located and available. And as for the third reason, additional command emphasis and oversight should be directed toward making and keeping reservations and ensuring that available quotas are used to the fullest extent possible. For those who cannot attend, prompt notification can permit other soldiers on wait-list status to use the vacant seat. Policies governing quota management are also important in dealing with quota utilization, such as expanded wait-listing (with better notification when wait-lists become reservations) and earlier reassignment of quotas from units that are not filling them to others that will.

In addition to these problems with quota utilization, our analyses also emphasize the importance of reducing the number of unqualified personnel, an area that lies outside traditional school system boundaries. The problem is one of both accuracy and size. The accuracy issue centers around forecasting of training requirements so as to ensure that school offerings better match the need. In our research, we found that reasonable current estimates could be developed using SIDPERS. This tool can also be used to make short-term forecasts, based on historical experience, of the number of soldiers needing training in the various career management fields (CMFs) and MOSs. Based on these findings, the Army organizations responsible for managing training requirements (i.e., ODCSPER, ARNG, and USARC) have begun to develop and apply such tools.

Finally, efforts should be taken to reduce the number of training requirements. The Army's "select-train-promote" policy is one example of how to reduce training requirements, in this case by limiting NCO education to only those NCOs who are selected for promotion. By extension, priorities could be established for determining which soldiers should be sent to reclassification training. Such training might be limited, for example, to soldiers with a remaining service obligation of a given duration, to those in selected MOSs, or as the Army now does, to those in high-priority units. Remaining soldiers might be sent to training as resources are available, or they might be qualified through other means (e.g., structured on-the-job training). Finally, incentives that reduce personnel turbulence can be especially effective, directed to minimizing attrition and job movements of MOS-qualified personnel to other positions for which they are not qualified.

ACKNOWLEDGMENTS

The authors benefited from support and assistance provided by many people in the U.S. Army. We owe particular thanks to our sponsors, Lieutenant General (ret.) Leonard D. Holder, Commanding General, U.S. Army Combined Arms Command, and Lieutenant General (ret.) John E. Miller, Deputy Commanding General, U.S. Army Training and Doctrine Command. In addition, key support and assistance was provided by the staff at the Army National Guard, Office of the Chief of the Army Reserve (OCAR), Office of the Deputy Chief of Staff for Operations (ODCSOPS), U.S. Army Forces Command (FORSCOM), First U.S. Army, the U.S. Army Reserve Command (USARC), and TRADOC's Office of the Deputy Chief of Staff for Training, including the TRADOC Coordinating Element (TCE) and Regional Coordinating Element (RCE) responsible for the Total Army School System. We also received key assistance from the staff of the State Adjutant Generals and Major U.S. Army Reserve Commands (MUSARCs) in the midwestern and southeastern United States—especially from the 108th Training Division of Charlotte, North Carolina, and from the staffs and instructors in the many Reserve Component Training Institutions (RCTIs) who bore the burden of data collection for this study. All provided access to data, information, and advice throughout the study, and they made it possible for us to observe training events and discuss training issues with staff, instructors, and students, all of which form the basis of the analysis in this report.

We are also grateful for assistance received from our RAND colleagues. Valuable contributions to this project were provided by Major (ret.) Gary Moody while he was a TRADOC Research Associate at RAND. Bruce Orvis and Bob Howe provided detailed and helpful reviews. Marilyn Yokota and Corazon Francisco provided invaluable research assistance. Jennifer Hawes-Dawson, Linda Daly, and Afshin Rastegar designed and managed data collection and processing.

We have benefited greatly from assistance provided by all these sources. Errors of fact or interpretation, of course, remain the authors' responsibility.

ACRONYMS

AC	Active Component
ADSW	Active Duty for Special Work
ADT	Active Duty for Training
AGR	Active Guard and Reserve
ANCOC	Advanced NCO Course
ARCOM	U.S. Army Reserve Command
ARNG	U.S. Army National Guard
ASI	Additional Skill Identifier
AT	Annual Training
ATRRS	Army Training Requirements and Resources System
ATSC	Army Training Support Center
BNCOC	Basic NCO Course
CA	Combat Arms
CAS3	Combined Arms Service and Staff School
CEAC	Army's Cost and Economic Analysis Center
CGSOC	Combined and General Staff Officers Course
CMF	Career Management Field
CONUS	Continental United States

CONUSA	Continental United States Army
CS	Combat Support
CSS	Combat Service Support
DAIG	Department of the Army Inspector General
DMDC	Defense Manpower Data Center
DMOSQ	Duty MOS Qualified
DoD	Department of Defense
DS/GS	Direct Support/General Support
DSL	Duty Skill Level
ECS	Equipment Concentration Site
FAST	Future Army Schools—Twenty One
FORSCOM	U.S. Army Forces Command
FY	Fiscal Year
GAO	General Accounting Office
GOSC	General Officer Steering Committee
IDT	Inactive Duty Training
IET	Initial Entry Training
MATES	Mobilization and Training Equipment Site
MDEP	Management Decision Package
MILEDC	Military Education Completed
MOA	Memorandum of Agreement
MOS	Military Occupational Specialty
MPRJ	Military Personnel Records Jacket
MTOE	Modified Table of Organization and Equipment
MUSARC	Major U.S. Army Reserve Command
NCO	Noncommissioned Officer
NCOA	NCO Academy

NCOES	NCO Education System
NGB	National Guard Bureau
O&M	Operations and Maintenance
O&S	Operations and Support
OES	Officer Education System
OPTEMPO	Operating Tempo
ORE	Operational Readiness Evaluation
PLDC	Primary Leadership Development Course
POI	Program of Instruction
POL	Petroleum, Oil, and Lubricants
RC	Reserve Component
RC3	Reserve Component Configured Courseware
RCCPDS	Reserve Component Consolidated Personnel Data System
RCE	Regional Coordinating Element
RCTI	Reserve Component Training Institution
RTS	Regional Training Site
RTS-I	Regional Training Site-Intelligence
RTS-M	Regional Training Site-Maintenance
RTS-Med	Regional Training Site-Medical
SIDPERS	Standard Installation/Division Personnel System
SL	Skill Level
SMA	State Military Academy
SQI	Skill Qualification Indentifier
SRC	Standard Requirements Code
SSSC	Self-Service Supply Center
TAG	The Adjutant General

TAPDB	Total Army Personnel Data Base
TASS	Total Army School System
TASSCA	Total Army School System Coordinating Activity
TATS	Total Army Training Structure
TCE	TRADOC Coordinating Element
TDA	Table of Distribution and Allowances
TDY	Temporary Duty
TOE	Table of Organization and Equipment
TRADOC	U.S. Army Training and Doctrine Command
USAR	U.S. Army Reserve
USARC	U.S. Army Reserve Command
USARF	U.S. Army Reserve Forces
UTES	Unit Training Equipment Site
VFAS	Vertical Force Accounting System

Chapter One

INTRODUCTION

BACKGROUND

For some time, the U.S. Army has recognized persistent problems in its extensive system of schools that provide technical and leadership training for the Reserve Components (RC), composed of the U.S. Army National Guard (ARNG) and the U.S. Army Reserve (USAR). Critics have suggested, for example, that in recent years the system lacks efficiency, provides an inconsistent quality of training, and is difficult to manage to meet the training needs of RC units.[1] To respond to these concerns, the Army initiated (beginning in FY94) a test of a "prototype" regional school system in the southeastern United States (Region C)—the states of North Carolina, South Carolina, Georgia, and Florida—with the intention of broadening it nationwide after a suitable period of testing. The prototype embodied significant changes to the organization and management of training, changes that were intended to raise standards and improve resource utilization. The changes also aimed at a longer-term goal—to establish a cohesive and efficient Total Army School System (TASS) of fully accredited and integrated schools to serve all Army components.

As the test got under way, the Army asked RAND Arroyo Center to analyze the operations of the system of schools serving the RC and assess whether the changes embodied in the prototype were improving the system's performance and efficiency. Initially, as the proto-

[1]See, for example, Department of the Army Inspector General (DAIG), *Special Assessment of Reserve Component Training*, Washington, D.C., January 11, 1993.

type school system was implemented (in FY94), the Arroyo Center published a "baseline" analysis assessing conditions and problems in Reserve Component Training Institutions (RCTIs) in three areas: training requirements and school delivery, training resources and costs, and quality of training.[2] The intention was to provide a "starting point" for measuring changes and improvements in the system and the prototype, to be followed by a subsequent assessment after a year of execution (in FY95).

OBJECTIVES

This report assesses the performance of the RC school system, based on measures used in the baseline assessment, in one of the key areas of assessment: training requirements and school delivery.[3] Principally, these measures quantify the size and nature of training requirements, the utilization of training capacity, and the delivery of training to the soldiers who need it. We examine the execution year of the prototype (FY95), compared to the baseline year (FY94), and we also examine the prototype in relation to the overall system. We analyze the impact of policy or structural changes that could enhance the training system's flexibility and effectiveness in meeting training requirements (i.e., by lowering turbulence to reduce demands on the system).

APPROACH

In assessing training requirements and school delivery in FY95, we took the same approach used in the baseline analysis. The research reported here focuses on the scope of the training requirements and the degree to which the RC schools meet the need—encompassing noncommissioned officer education classes for NCOs (NCOES) and reclassification training of personnel who are not qualified in their duty military occupational specialty (MOS) to make them duty MOS

---

[2]See John D. Winkler et al., *Assessing the Performance of the Army Reserve Components School System*, Santa Monica, CA: RAND, MR-590-A, 1996, for a complete discussion of the baseline assessment.

[3]A companion report examines resource use and efficiency of training. See Michael G. Shanley, John D. Winkler, and Paul S. Steinberg, *Resources, Costs, and Efficiency of Training in the Total Army School System*, Santa Monica, CA: RAND, MR-844-A, 1997.

qualified (DMOSQ)—for U.S. Army reservists and guardsmen. In terms of school production (capacity), the research compares the DMOSQ and NCOES training requirements against the capacity of RC schools to meet them, analyzing how the school system utilizes capacity and produces trained graduates. Finally, we examine whether the soldiers receiving training are the ones who need it.

We used a number of data sources to conduct these analyses. To estimate DMOSQ and NCO training requirements, we used national-level Army personnel, training, and force structure data. Specifically, we used (as we did in the previous analysis) the Army National Guard's (ARNG's) and U.S. Army Reserve's (USAR's) Standard Installation/Division Personnel System (SIDPERS); in this case, we used data from FY93–FY95. As part of this analysis, and to analyze training capacity, we used national-level data from the Army Training Requirements and Resources System (ATRRS), which provided school-level data on course offerings, quotas, inputs, and graduates and individual-level data on course attendees; in this case, we used FY94 and FY95 data. In addition, in the research reported here, we had our SIDPERS-based estimates of training requirements validated by First U.S. Army, which conducted Operational Readiness Evaluations (OREs) of 18 units and 1,300 soldiers. See Appendix A for the results of the ORE analysis.

ORGANIZATION OF THIS REPORT

In Chapter Two, we examine training requirements and school delivery for NCOES. Chapter Three examines training requirements and school delivery for reclassification (DMOSQ) training. Chapter Four presents the results of a modeling effort showing how reducing turbulence affects the level of the DMOSQ training requirement. Chapter Five presents our conclusions and recommendations.

Appendix A discusses the results of the ORE analysis mentioned above, and Appendix B describes the data used to track training requirements and school delivery.

Chapter Two

NCOES TRAINING REQUIREMENTS AND SCHOOL DELIVERY

In this chapter we examine training requirements and school delivery for NCO education, looking first at the national results in FY95 and examining trends from FY94 to FY95 at the national level. We then examine results in Region C and compare them to the national results during the period of observation. We begin by briefly discussing the courses that make up the NCOES for soldiers in grades E-4 through E-6.

COURSES THAT MAKE UP NCOES TRAINING

There are three primary NCO professional development courses:

- **Primary Leadership Development Course (PLDC).** This two-week resident course provides training in leadership techniques, Army training methods, and doctrine (not MOS-specific). Completion is required for promotion to E-5.
- **Basic NCO Course (BNCOC).** This two-phase course introduces basic skills for NCOs at grades E-6 and above. Phase 1 provides general leadership training that is not MOS-specific, which can be taught in either six consecutive days or three weekends. Phase 2 contains MOS-specific material usually developed by the proponent schools and taught in two-week resident mode for most MOSs. (Some MOSs take longer.) Completion is a requirement for promotion to E-6.
- **Advanced NCO Course (ANCOC).** Like BNCOC, ANCOC is a two-phase course on becoming an effective platoon sergeant or senior section sergeant (E-7). Phase 1 is common leader training

that is not MOS-specific and includes a field training exercise. It can be taught in either fourteen consecutive days or seven weekends. Phase 2 contains MOS-specific material usually developed by the proponent schools and is taught in a two-week resident mode for most MOSs. (Some MOSs take longer.) Completion is a requirement for promotion to E-7.

THE NCOES TRAINING REQUIREMENT REMAINS SIZABLE IN FY95

As our "baseline" analysis showed, a large number of NCOs appear to need NCOES education, according to Army personnel records. This number includes soldiers who are selected for promotion to the next-higher grade and need to complete required NCOES courses. Another group comprises soldiers who have already been promoted but have not fully completed the NCOES required for their grade.[1]

Soldiers Selected for Promotion and Needing Training

Table 2.1 shows the number of soldiers in grades E-4 through E-6 who were promoted in FY95 across the nation. Overall promotion rates for soldiers in grades E-4 through E-6 were about 8 percent in FY95. Thus, based on FY95 promotions, NCOES requirements would be as follows—ARNG, 14,163 and USAR, 7,530—for a total of 21,693, distributed across grades as shown in the table.

Table 2.2 shows the educational status of NCOs promoted in FY95. The table's left side shows the number of NCOs promoted from that grade in FY95 (as shown in Table 2.1) and, of these, the number and percentage who completed the appropriate NCOES in the year before their promotion (FY94) and the number and percentage who still needed NCOES at the start of FY95. The right side of Table 2.2 shows the numbers and percentages of NCOs who completed the appropriate NCOES in FY95, along with the remaining NCOs who still need NCOES at the end of the year in which they were promoted.

[1] Some of these soldiers may have been promoted before specific NCOES courses were required for promotion. Others may have partially but not fully completed the required NCOES courses.

Table 2.1

FY95 E-4 through E-6 Promotions by Component and Grade

Component/ Grade	Number on Hand 9/30/94	Number Promoted 9/30/95	Percent Promoted 9/30/95
ARNG			
E-4	82,692	7,496	9.1
E-5	65,714	4,416	6.7
E-6	34,099	2,251	6.6
Total	182,505	14,163	7.8
USAR			
E-4	47,363	4,172	8.8
E-5	25,669	2,162	8.4
E-6	19,759	1,196	6.1
Total	92,791	7,530	8.1
Both			
E-4	130,005	11,668	9.0
E-5	91,383	6,578	7.2
E-6	53,858	3,447	6.4
Total	275,246	21,693	7.9

SOURCE: ARNG and USAR SIDPERS, November 1994, November 1995.

As can be seen in the table, nearly half the NCOs promoted in FY95 completed their NCOES in the year before their promotion. The other half (about 11,000 NCOs) required training in FY95. Many of these soldiers completed training during the year of promotion. By the end of FY95, however, about one-fifth of the NCOs in grades E-4 through E-5 (about 4,500 NCOs) still had not fully completed their NCO training.

These data indicate that the Army was implementing its policy of "select-train-promote" for the majority of soldiers but not yet to the needed extent. In FY95, about 79 percent of soldiers in grades E-4 through E-6 completed the required NCOES in the year they were promoted or in the year before. In FY94, the percentage was about 78 percent.

Furthermore, "select-train-promote" was implemented to a greater extent in the ARNG than in the USAR for grades E-4 to E-5 over the two-year period. However, as shown in Figure 2.1, the USAR made

Table 2.2

Promotions and Completions of NCOES by Component and Grade

Component/ Grade	Number Promoted in FY95	Number Received NCOES in FY94 *(Percent)*	Number Needing NCOES in FY95 *(Percent)*	Number Received NCOES in FY95 *(Percent)*	Number Needing NCOES *(Percent)*
ARNG					
E-4	7,496	4,105	3,391	2,834	557
		(54.8)	*(45.2)*	(37.8)	(7.4)
E-5	4,416	1,966	2,450	1,212	1,238
		(44.5)	*(55.5)*	*(27.5)*	*(28.0)*
E-6	2,251	1,124	1,127	556	571
		(49.9)	*(50.1)*	*(24.7)*	*(25.4)*
Total	14,163	7,195	6,968	4,602	2,366
		(50.8)	*(49.2)*	*(32.5)*	*(16.7)*
USAR					
E-4	4,172	2,094	2,078	1,136	942
		(50.2)	*(49.8)*	*(27.2)*	*(22.6)*
E-5	2,162	817	1,345	505	840
		(37.8)	*(62.2)*	*(23.4)*	*(38.9)*
E-6	1,196	533	663	342	321
		(44.6)	*(55.4)*	*(28.6)*	*(26.8)*
Total	7,530	3,444	4,086	1,983	2,103
		(45.7)	*(54.3)*	*(26.3)*	*(27.9)*
Both					
E-4	11,668	6,199	5,469	3,970	1,499
		(53.1)	*(46.9)*	*(34.0)*	*(12.8)*
E-5	6,578	2,783	3,795	1,717	2,078
		(42.3)	*(57.7)*	*(26.1)*	*(31.6)*
E-6	3,447	1,657	1,790	898	892
		(48.1)	*(51.9)*	*(26.1)*	*(25.9)*
Total	21,693	10,639	11,054	6,585	4,469
		(49.0)	*(51.0)*	*(30.4)*	*(20.6)*

SOURCE: ARNG and USAR SIDPERS, November 1994, November 1995.

progress in implementing this policy (moving from 58 percent in FY94 to 72 percent in FY95, while the ARNG fell back somewhat (going from 90 percent in FY94 to 83 percent in FY95).

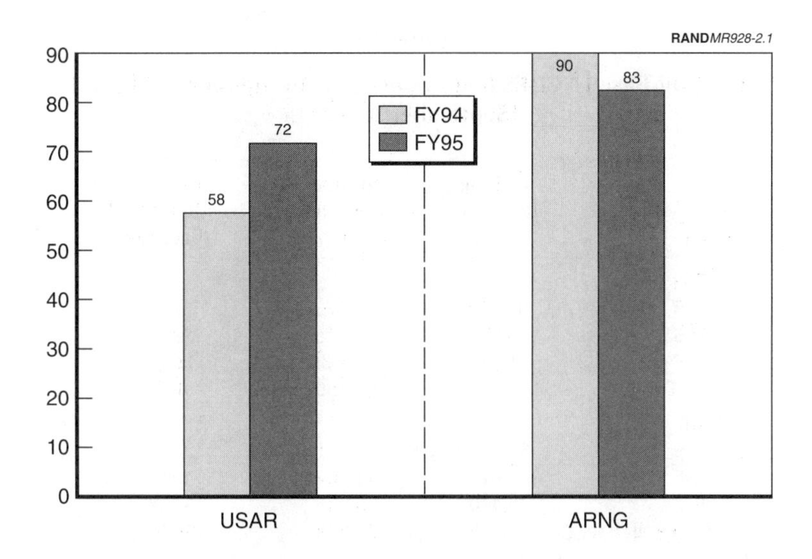

Figure 2.1—Percent of Promotees Trained, by End of Year Promoted

Soldiers Already Promoted But Still Requiring Training

We now turn to the other group of NCOs who require NCO education—NCOs who have been promoted but have not fully completed the required NCOES courses. As shown in Table 2.3, over 79,000 NCOs (38.3 percent of on-hand personnel) fell in this category at the start of FY95. This, however, was an improvement over FY94, when 94,450 NCOs (43.8 percent of on-hand personnel) had not fully completed the NCO courses required for their grade at the start of the year.

These nonqualified NCOs, though seemingly numerous, may not all constitute a training requirement, because some leave the force during the year (hence, there may be little point in training them). If we exclude NCOs who attrit during FY95, the "backlog" of unqualified NCOs in grades E-5 through E-7 who required training was less than 79,000 but still large. During FY95, for example, of the roughly 79,000 NCOs in grades E-5 through E-7 who began the year non-

Table 2.3

Backlog-Based NCOES Requirements by Component and Grade (September 30, 1994)

Component/ Grade	Drilling Reservist Number	Number Needing NCOES (backlog)	Percent Needing NCOES (backlog)
ARNG			
E-5	81,932	30,773	37.6
E-6	40,792	15,648	38.4
E-7	11,749	3,159	26.9
Total	134,473	49,580	36.9
USAR			
E-5	32,872	12,481	38.0
E-6	24,017	11,762	49.0
E-7	15,440	5,355	34.7
Total	72,329	29,598	40.9
Both			
E-5	114,804	43,254	37.7
E-6	64,809	27,410	42.3
E-7	27,189	8,514	31.3
Total	206,802	79,178	38.3

SOURCE: ARNG and USAR SIDPERS, November 1994.

qualified, 19,000 left the force.[2] Hence the training requirement for FY95 consisted of approximately 60,000 soldiers (the unqualified NCOs who remained in the force throughout the fiscal year).[3]

Table 2.4 shows, for NCOs requiring training and remaining in the force throughout FY95, our estimate of the total "true" FY95 NCOES

[2]A handful of the unqualified NCOs were promoted, with many receiving NCOES for their new, higher grade.

[3]The "true" training requirement might be somewhat higher because some of the NCOs who left the force may have done so for lack of training opportunity. Had training been available, they might have stayed and hence the number of NCOs requiring training would be larger than 60,000.

Table 2.4

FY95 NCOES Requirements by Component and Course

Component/ Course	Number Promoted in FY95	Number Needing Course in FY95	Number Needing Course for Grade FY95 (backlog)	Total Needing Course FY95
ARNG				
PLDC	7,496	3,391	23,238	26,629
BNCOC	4,416	2,450	12,331	14,781
ANCOC	2,251	1,127	2,492	3,619
Total	14,163	6,968	38,061	45,029
USAR				
PLDC	4,172	2,078	8,636	10,714
BNCOC	2,162	1,345	9,333	10,678
ANCOC	1,196	663	4,161	4,824
Total	7,530	4,086	22,130	26,216
Both				
PLDC	11,668	5,469	31,874	37,343
BNCOC	6,578	3,795	21,664	25,459
ANCOC	3,447	1,790	6,653	8,443
Total	21,693	11,054	60,191	71,245

SOURCE: ARNG and USAR SIDPERS, November 1994, November 1995.

training requirement for specific courses (PLDC, BNCOC, and ANCOC).[4] It shows the number of personnel needing each course, in each component and category of training requirement (promotion versus backlog). Hence, our best estimate of the "true" NCOES training requirement in FY95 was about 71,000 (about 45,000 in the ARNG as compared with nearly 26,000 in the USAR), including 11,000 newly promoted and 60,000 previously promoted NCOs who remained in the force throughout the year.

These numbers, though large, represent a considerable improvement over FY94. Taking attrition in each year into account, "true"

[4]For example, the table determines requirements for PLDC based on E-4s who were promoted in FY95 and who had not completed PLDC, plus soldiers in grades E-5 who have not completed PLDC and remained in the force at that grade throughout FY95 (according to SIDPERS records). Eligibility for BNCOC is calculated based on promoted E-5s and nonqualified E-6s, while ANCOC is calculated based on promoted E-6s and nonqualified E-7s).

NCOES training requirements fell to 71,245 for FY95 (shown in Table 2.4) from 84,573 in FY94 (52,604 in the ARNG and 31,969 in the USAR).[5] We estimated the total training requirements by course in FY94 and FY95 as follows: PLDC—44,693 and 37,343; BNCOC—30,055 and 25,459; ANCOC—9,825 and 8,443.

Overall this represents a reduction of 15.8 percent in the number of nonqualified NCOs from FY94 to FY95. These reductions in NCO requirements largely occur for two reasons. First, as shown earlier, the ARNG and USAR continued to implement their "select-train-promote" policy, which emphasizes that only soldiers selected for promotion should be sent to NCOES. Hence the backlog of untrained NCOs is remaining constant (at worst) or decreasing. In addition, the promoted but untrained NCOs began to disappear—they either left the force, received the training needed for their current grade, or received training before being promoted to the next-higher grade.[6] If these trends continue, the NCOES training requirement should continue to fall in the coming years in the direction of its "steady state," where the annual training requirement would be driven by the annual promotion rate.[7]

DEMAND FOR NCO TRAINING STILL EXCEEDS SUPPLY, BUT TO A LESSER EXTENT

For purposes of our analyses, we are interested in relating NCOES training requirements to the capacity of training institutions to meet them. As in our baseline assessment, we compare training requirements against the capacity available in RC schools to deliver training,

---

[5]Our baseline report (Winkler et al., 1996) did not take attrition into account in calculating the potential NCOES training requirement. Hence the FY94 numbers shown here are smaller than those shown in the earlier report (84,573 versus 104,417).

[6]A reduction in the size of the U.S. Army Reserve Components was also under way during this time. The number of drilling reservists fell from 514,025 in FY94 to 489,970 in FY95—a reduction of 4.7 percent. Hence, the decrease in unqualified NCOs (15.8 percent) was disproportionate to the reduction in force size that occurred during this period.

[7]These trends appear to be continuing. For example, during FY95, out of the 60,000 unqualified NCOs, about 10,000 completed the required NCOES course, leaving 50,000 NCOs in the RC who appear to still require NCO education at the beginning of FY96 (some of whom will attrit).

as well as the utilization of that capacity. The measure of capacity is the number of programmed training seats (quotas), and the measure of capacity utilization is the percentage of quotas filled by students.

Figure 2.2 compares NCO training requirements against the total quotas established at the start of the year for PLDC, BNCOC, and ANCOC,[8] for the two fiscal years covered in the assessment. For sizing purposes, we combine the promotion-based and backlog-based requirements into one estimate (71,245 soldiers in FY95).

As indicated in our baseline assessment, the RC school system does not have sufficient capacity to meet its total NCOES training requirements, when promotion and backlog-based requirements are considered together. This observation held in FY95, as the number of NCOES quotas available in RC schools (44,568 seats) still fell short of the 71,245 training requirement.

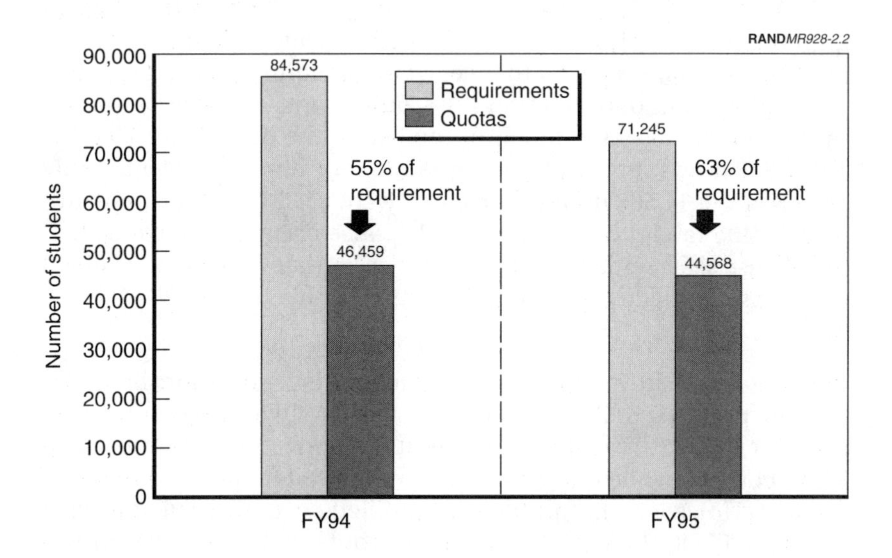

Figure 2.2—Training Demand and Supply for NCOES Training Are Better Aligned in FY95 Than in FY94

---

[8]BNCOC and ANCOC quotas are for the MOS-specific (Phase 2) portion of these courses.

However, as shown in Figure 2.2, schools were better able to meet training requirements in FY95, as compared to FY94, primarily because the NCOES requirement decreased. In FY94, there were about 84,573 NCOs in need of training, whereas in FY95, this number fell to 71,245. Meanwhile, the schools experienced a drop in the number of NCOES quotas in FY95 compared to FY94, but they fell to a lesser extent than the requirement. Specifically, while quotas fell for PLDC from 25,841 to 25,233 and from 20,618 to 19,335 for NCOES Phase 2 (ANCOC and BNCOC), these quotas comprised 63 percent of requirement in FY95 versus 55 percent of requirement in FY94. These quotas were roughly double the number of promotions in FY95; this suggests that if NCOES training requirements can approach the "steady state," less capacity will be needed to meet the requirement.

NCOES TRAINING CAPACITY IS NOT FULLY UTILIZED

RAND Arroyo Center's baseline assessment of NCOES training requirements and school delivery showed considerable inefficiencies in using available capacity to meet training requirements. Despite a better alignment between training requirements and school quotas, these inefficiencies remained in the use of the quotas in FY95. As Table 2.5 shows, many of these quotas went unused. Overall, only 28,163 of the 44,568 quota allocations were used (about 63 percent), with 16,405 (about 37 percent) lost because classes were cancelled[9] or because seats remained unfilled in classes that were held. Unfilled training seats made up most of the unused quotas (about 28 of the 37 percent).

The table also shows that these problems were worse for the MOS-specific portions of ANCOC and BNCOC (NCOES Phase 2) courses than for PLDC: Two-thirds of the quota allocations were used in PLDC courses, while only three-fifths were used in the NCOES Phase 2 ones. However, the problem of unfilled seats was substantially greater in PLDC than in NCOES Phase 2 courses: The vast majority of

---

[9]Classes shown as cancelled combine two potential occurrences. The first is when the decision not to hold it was made after the class is scheduled to report (termed "cancelled" in the ATRRS system), as well as ones where the decision not to hold it was made before the class was scheduled to report (termed "nonconducted" in the ATRRS system).

Table 2.5

NCOES Quota Allocations by Course in FY95

Course	Number Needing NCOES	Quota Allocations *(Percent of No. Needing)*	Quotas Cancelled *(Percent of Allocation)*	Quotas Not Filled *(Percent of Allocation)*	Total Quotas Used *(Percent of Allocation)*
PLDC	37,343	25,233	160	8,301	16,772
		(68)	*(1)*	*(33)*	*(66)*
NCOES Phase 2 (ANCOC, BNCOC)	33,902	19,335	3,605	4,339	11,391
		(57)	*(19)*	*(22)*	*(59)*
Total	71,245	44,568	3,765	12,640	28,163
		(63)	*(9)*	*(28)*	*(63)*

SOURCE: ATTRS School Aggregate file November 1995.

lost PLDC quotas were attributable to unfilled seats, while in ANCOC/BNCOC Phase 2, they are attributable in nearly equal part to cancelled courses and unfilled seats.[10]

From what we could observe from our interviews and observations, there are a number of reasons for seats going unfilled. Many units are still not using the reservation system (ATRRS), some soldiers could not get orders, and other soldiers cancelled reservations or did not show up for training and suitable replacements could not be found in time.[11] In addition, anecdotal evidence suggests that heightened standards produced a higher number of turnbacks (e.g., through inability to meet Army physical fitness standards). All combined to create "lost capacity" and wasted resources, which is particularly troublesome given that requirements exceed capacity for these courses.

---

[10]We also examined capacity utilization in the "common leader" portion of ANCOC and BNCOC courses (so-called Phase 1 courses). Here the quota fill rate is quite good, amounting to 92 percent of quotas in FY95. Most lost quotas were due to cancelled courses.

[11]For example, we observed that in FY95, reservations were made for less than half of the quotas available in NCOES Phase 2 classes (9,216 reservations versus 19,335 quotas). Some soldiers holding reservations did not show up (a particular problem in PLDC), while other attendees "walked on" and attended without reservations.

Inefficiencies in Quota Use Grew Worse Across the Fiscal Years

When we examine how well the RC school system did in meeting the NCOES training requirements in FY95, we see that quota utilization actually grew worse in FY95 as compared to FY94. As shown in Figure 2.3, while the percentage of cancelled quotas was reduced by one percentage point for PLDC courses in FY95, this was overshadowed by an increase in unfilled seats of 11 percentage points. In NCOES Phase 2, the percentage of both cancelled and unfilled quotas increased slightly—by 3 and 2 percentage points, respectively. Altogether, the quota utilization in PLDC and NCOES Phase 2 declined by 10 percentage points and 5 percentage points, respectively.[12]

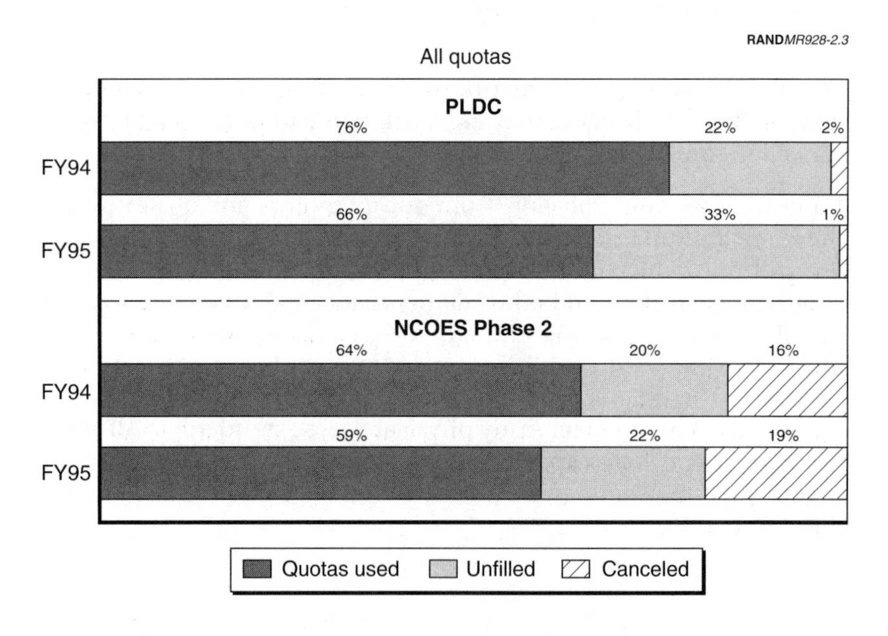

Figure 2.3—The Use of Quotas in NCOES Training Grew Worse from FY94 to FY95

---

[12]The fill rate in ANCOC/BNCOC Phase 1 courses improved across the fiscal years, rising to 92 percent from 89 percent, primarily because of a reduction in the number of cancelled courses.

Some Students Attending NCO Courses Do Not Require Training

There is another problem with the utilization of school capacity: Some soldiers taking NCOES courses do not appear to need this training. As indicated earlier, under the "select-train-promote" policy, only those soldiers selected for promotion should be sent to the NCO course required for the next-higher grade. However, when we observe the completion of military education (as reported in SIDPERS) in relation to soldiers' grades, we see that many soldiers attending NCOES were not subsequently promoted (hence, they may not have been selected for promotion and thus were not eligible to attend).

Based on Army personnel data, soldiers were sent to NCOES courses at two to three times the rate soldiers in the grade were being promoted. As shown in Table 2.6, among ARNG and USAR E-4s in FY95, 16 percent had completed PLDC at some point in their career, but only 9 percent of E-4s were promoted to grade E-5 in FY95. This "overtraining" is even more extensive among E-5s and E-6s, who completed BNCOC or ANCOC, respectively, at two to three times the rate at which soldiers in these grades were promoted in FY95.

This problem can also be seen when viewed from a different perspective, as shown in Table 2.7. We examined ATRRS class rosters of PLDC and NCOES Phase 2 courses and determined whether attendees of a given grade were promoted prior to or within a year of attending the course. The table shows, for example, that among all

Table 2.6

Number of NCOs Being Trained and Being Promoted

Grade (Course)	Received Course (Percent)	Promoted Annually (Percent)
E-4 (PLDC)	16	9
E-5 (BNCOC 2)	16	7
E-6 (ANCOC 2)	18	6

SOURCE: ARNG and USAR SIDPERS, November 1995.

Table 2.7

Number of NCOs Attending School Promoted Within a Year

Soldiers Attending Course in FY94	Percent Promoted as of	
	FY94	FY95
E-4s attending PLDC	17	47
E-5s attending BNCOC 2	17	50
E-6s attending ANCOC 2	12	48

SOURCE: ATTRS School Aggregate file November 1994; ARNG and USAR SIDPERS, November 1994, November 1995.

E-4s attending PLDC in FY94, less than half were promoted to E-5 by the end of FY95. The results are similar for E-5s and E-6s attending Phase 2 classes in BNCOC and ANCOC, respectively.

Together, these results illustrate a difficult problem confronting the TASS related to its performance and efficiency. The system faces problems both in using available capacity *and* in ensuring that the "right" NCOs are sent to school. Moreover, since some graduates of NCOES courses are not being promoted, the portion of the NCOES requirement being met is even smaller than indicated by simple graduation numbers, as discussed below.

SCHOOL SYSTEM PRODUCTION IS COMPARABLE ACROSS THE FISCAL YEARS

We now examine the "production" of fully qualified NCOs in the RC school system, again drawing on the measures used in our baseline assessment:

- The number of graduates and the graduation rate;
- The ratio of graduates to quota allocations, which shows the degree to which production met initial capacity;
- The ratio of graduates to requirements, which compares production with the overall need for NCOES training.

These measures are shown in Table 2.8. For PLDC, the ratio of graduates to quotas was five percentage points lower than that of the NCOES Phase 2 courses, but it had a substantially lower graduation rate (20 percentage points). The NCOES Phase 2 courses had a very high graduation rate (96 percent), but since so much initial capacity went unused, the ratio of graduates to the requirement is about the same as PLDC (.32 versus .35).[13] Ultimately, the 23,691 graduates of all of these courses represent only 33 percent of the overall requirement shown in Table 2.5.[14]

Figure 2.4 summarizes and compares school capacity utilization and production across the fiscal years examined in the study. As Figure 2.2 showed, the school system did better in terms of quota allocations versus requirements in FY95. Even though the total number of quotas fell from FY94 to FY95, the total requirements fell faster between the two fiscal years; thus, quotas were 63 percent of requirements in FY95 versus 55 percent in FY94. As Figure 2.3 showed, the use of quota allocations was worse in FY95 than it was in FY94. This

Table 2.8

NCOES Production by Course in FY95

Course	Quota Allocations	Inputs	Grads	Graduation Rate	Ratio of Grads to Quotas	Ratio of Grads to Req'ts
PLDC	25,233	16,772	12,795	.76	.51	.35
NCOES Phase 2	19,335	11,391	10,896	.96	.56	.32
Total	44,568	28,163	23,691	.84	.53	.33

SOURCE: ATTRS School Aggregate file November 1995.

[13]This comparison is done for purposes of sizing output in relation to training demand. Since, as shown earlier, some graduates of NCOES courses do not appear to require training, it does not accurately measure the amount of the "true" requirement met by the system. It shows, however, that the supply of qualified NCOs is considerably smaller than the need.

[14]Given better quota utilization, ANCOC/BNCOC Phase 1 courses show more favorable measures of production, providing 16,281 graduates (82 percent of quotas), despite a lower graduation rate than seen in NCOES Phase 2 (83 percent versus 96 percent).

is captured in Figure 2.4, where we see that 71 percent of the quotas in FY94 were filled (i.e., with student inputs) versus 63 percent in FY95. However, as Figure 2.4 also shows, the school system also did a better job in FY94 than in FY95 of producing graduates from these inputs—89 percent versus 84 percent. Hence the output of the system in relation to its need was about the same in FY95 as in the previous year. Specifically, graduates represented 35 percent of the requirements in FY94 versus 33 percent in FY95.[15]

The story is similar in PLDC and NCOES Phase 2 courses: PLDC graduates represented 51 percent of the quota allocations in FY95 versus 65 percent in FY94. In NCOES Phase 2, graduates represented 56 percent of quota allocations in FY95 versus 62 percent in FY94. Ultimately, PLDC graduates represented 34 percent of the requirement, while NCOES Phase 2 graduates represented 32 percent. This

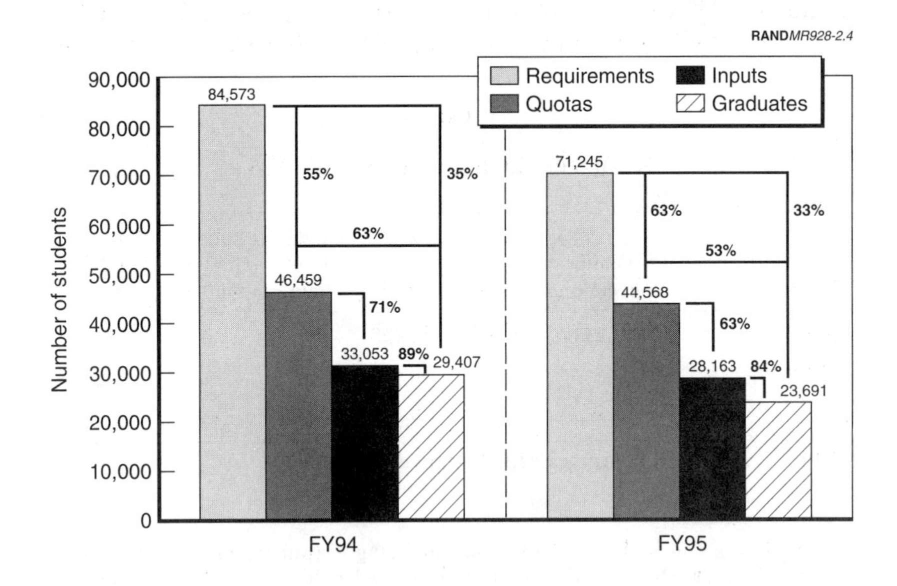

Figure 2.4—The System Did About the Same in Producing NCOES Graduates Across the Fiscal Years

[15]Again, this shows the overall output in relationship to the overall requirement, but since some graduates did not appear to need this training, the percentage of the "true" requirement that was met was smaller in both fiscal years.

compares with 37 percent and 32 percent in FY94 for PLDC and NCOES Phase 2 courses, respectively.

PERFORMANCE IN REGION C IS EQUIVALENT TO OTHER REGIONS

In the face of these trends in school capacity utilization and delivery for Army NCO education, it is perhaps reassuring to observe that performance of the training system in the prototype was, on the whole, equal to the rest of the nation on most of these measures. For example, with respect to the size and nature of NCOES training requirements, Region C performed equally well as other regions in implementing "select-train-promote" and better in reducing the "backlog" of nonqualified NCOs. For example, during FY95 in Region C, 80 percent of RC soldiers promoted from grades E-4 through E-6 received the required NCOES prior to or during the year of promotion (versus 79 percent in other regions).[16] In Region C, 73 percent of NCOs in grades E-5 through E-7 are shown as fully qualified by the end of FY95, compared to 67 percent in other regions.[17]

In viewing school system capacity and its utilization, it must be kept in mind that geographic flows of students occur between regions of the country and, indeed, consolidation of training locations to a smaller number of national and regional sites is an aim of the TASS.[18] Hence, in examining school system performance in Region C, the key measures of merit are quota use and output of fully qualified NCOs compared to *quotas* established in Region C schools. Sizing the output of Region C schools as compared to the specific Region C *requirement* is misleading, since some soldiers who reside in Region C will take their schooling outside Region C, and some of the students attending schools in Region C come from other parts of the country.

---

[16]In FY94, these figures were 78 percent in both Region C and other regions.

[17]In FY94, these figures were 69 percent and 62 percent in Region C and other regions, respectively.

[18]For example, the ARNG is seeking to reduce the number of sites where combat arms training is conducted. Currently, for example, most annual training of M1A1 armor crewmen occurs at Gowan Field, Idaho. Similar initiatives are under way in other functional areas.

When we view quota utilization in Region C schools, we see they did about the same as the rest of the nation. Table 2.9 shows that in Region C in FY95, total quota use in NCOES courses was about the same overall (65 percent versus 63 percent) and nearly identical percentagewise to the figures at the course level in the rest of the nation. Again, we see problems with cancelled courses and unfilled seats in Region C and elsewhere, with unfilled seats the primary source of lost capacity in PLDC. Meanwhile, there was more of a balance between cancelled and unfilled quotas in NCOES Phase 2.

Over the time periods examined in the study, Region C schools performed about the same as schools in the other parts of the nation. Quota utilization for PLDC fell to 67 percent in FY95 from 72 percent in FY94 (when there were 3,620 inputs versus 5,015 allocations), while quota use in the rest of the nation fell to 66 percent in FY95

Table 2.9

NCOES Quota Allocations by Course by Region in FY95

Region/ Course	Quota Allocations	Quotas Cancelled *(Percent of Allocation)*	Quotas Unfilled *(Percent of Allocation)*	Total Quotas Used *(Percent of Allocation)*
Region C				
PLDC	3,824	0	1,268	2,556
		(0)	*(33)*	*(67)*
NCOES Phase 2	1,394	320	259	815
		(23)	*(19)*	*(58)*
Total	5,218	320	1,527	3,371
		(6)	*(29)*	*(65)*
Other				
PLDC	21,409	160	7,033	4,216
		(1)	*(33)*	*(66)*
NCOES Phase 2	17,941	3,258	4,080	10,576
		(18)	*(23)*	*(59)*
Total	39,350	3,418	11,113	24,792
		(9)	*(28)*	*(63)*
All				
Total	44,568	3,765	12,640	28,163
		(9)	*(28)*	*(63)*

SOURCE: ATTRS School Aggregate file November 1995.

from 72 percent in FY94 (when there were 16,239 inputs versus 20,826 allocations). Meanwhile, quota use in Region C NCOES Phase 2 courses rose to 58 percent of allocations in FY95 from 49 percent of allocations in FY94. In the rest of the nation, quota use in NCOES Phase 2 courses fell to 59 percent in FY95 from 65 percent in FY94 (when there were 12,248 inputs as compared to 18,705 Phase 2 quotas).

A noteworthy change in Region C was a reduction in cancelled courses. Specifically, in PLDC courses, the percent of quotas lost from cancelled courses fell from 9 percent in FY94 to 0 percent in FY95 (while remaining stable at about 1 percent in the rest of the nation). In terms of NCOES Phase 2 courses, the percent of cancelled courses actually *fell* in Region C from 33 percent to 23 percent between the two fiscal years (compared to a rise from 14 percent to 18 percent elsewhere). However, unfilled quotas remain a problem, particularly in PLDC.

Table 2.10 shows that production rates in Region C were about the same as the rest of the nation in FY95. In Region C, the overall graduation rate from NCOES courses was about five percentage points lower than in other regions, with most of the difference attributable to its lower proportion of NCOES Phase 2 allocations (versus PLDC).[19] Graduation rates and output for specific NCOES courses were quite similar.

IMPLICATIONS

The findings presented in this chapter indicate that considerable problems remain with respect to the management of Reserve Com-

---

[19]One area in which Region C experienced problems with capacity use and production was in ANCOC/BNCOC Phase 1 (common leader training) courses. In FY95, 65 percent of initial quotas were used, compared to 96 percent in the rest of the nation. The graduation rate was also considerably lower in FY95, amounting to 73 percent of attendees (compared to 84 percent in the rest of the nation). Moreover, the quota fill rate worsened in Region C in FY95, while increasing elsewhere. We are not entirely sure of the reasons for these problems. We note, however, a large number of unprogrammed, "walk-on" attendees in Phase 1 courses in the other regions (inputs actually exceeded quotas in courses that were held). In addition, anecdotal reports suggested that there were too few sites offering NCO Phase 1 courses in Region C, making it more difficult to resource and support soldiers to attend this training.

Table 2.10

NCOES Production by Course by Region in FY95

Region/ Course	Quota Allocations	Inputs	Graduates	Graduation Rate	Ratio of Grads to Quotas
Region C					
PLDC	3,824	2,556	1,911	.75	.50
NCOES Phase 2	1,394	815	771	.95	.55
Total	5,218	3,371	2,682	.80	.51
Others					
PLDC	21,409	14,216	10,884	.77	.51
NCOES Phase 2	17,941	10,576	10,125	.96	.56
Total	39,350	24,792	21,009	.85	.53
All					
Total	44,568	28,163	23,691	.84	.53

SOURCE: ATTRS School Aggregate file November 1995.

ponent NCO training requirements and with the use of school capacity to meet these requirements. The primary problems lie with the size of the requirement (especially in relation to supply of training), the efficient use of available training capacity, and course attendance by qualified students.

With these observations in mind, we see a continuing need for policies and incentives that reduce NCOES training requirements and improve quota use while ensuring the "right" soldiers are sent to NCO courses. The trends we have observed suggest that the policy of "select-train-promote" still needs to be fully implemented, and actions must be taken to reduce the "backlog" of unqualified NCOs. In addition, policies are needed to ensure that available capacity is better utilized (e.g., by creating incentives for using the reservation system appropriately). With continued emphasis on training only those soldiers who require NCOES, and as unqualified NCOs are trained or leave the force, the training requirement should decline until it approaches "steady state," allowing for continued consolidation of NCOES training.

But without further steps to improve the operation of the school system, a reduction in training requirements in the future can exacerbate the problems of training delivery shown earlier, because soldiers needing NCOES will be fewer in number and more dispersed. This can make it even harder to schedule and fill NCOES courses for only those soldiers who really need them. Hence the Army will need to place greater emphasis on using the reservation system and getting appropriate soldiers to the training. Further, economies of scale will argue for further consolidation of AT so that scheduled classes are more likely to be filled. Moreover, there may be MOSs where the number of RC NCOs is so small that training cannot be effectively managed within the RC system. Here, the AC may need to fill voids, perhaps by conducting more training of RC NCOs than is currently the case.

Chapter Three

RECLASSIFICATION TRAINING REQUIREMENTS AND SCHOOL DELIVERY

This chapter addresses requirements for reclassification (DMOSQ) training, along with school capacity utilization and production of qualified graduates. As in the previous chapter, we look first at national results in FY95, for the system as a whole, and as compared to FY94. We then examine results in Region C and compare them with national results during the period of observation.

THERE IS A SIZABLE DMOSQ TRAINING REQUIREMENT IN FY95

As in our earlier report, these analyses cover all drilling guardsmen and reservists in grades E-1 through E-9. We begin by noting the number of drilling reservists who are qualified to hold their duty MOS.[1] Among those not qualified for their current position, we further note the number of soldiers who are new entrants to the military, requiring initial entry training (IET), as compared to soldiers with prior military service who require reclassification training.[2]

[1]The methodology we used to estimate reclassification training requirements is described in Appendix B. Briefly, we use SIDPERS records from the ARNG and USAR at the start of the fiscal year (here 1995) to estimate the number of soldiers qualified for their duty MOS and those not qualified in each Reserve Component. These numbers provide a snapshot of the training requirement at the start of the fiscal year, and are shown in relation to the number of training seats available during the fiscal year, to provide an overall comparison of training need versus training supply.

[2]We define soldiers as DMOSQ or needing reclassification training by matching their duty MOS through the first three digits against the first three digits of the primary, secondary, or additional MOS, as shown in SIDPERS records. If no such match exists, we define the individual as non-DMOSQ and hence in need of reclassification training.

Subsequently, these figures are viewed against the capacity of RC schools to conduct reclassification training and efficiency in using this capacity.[3]

Table 3.1 shows the number of soldiers shown in SIPDERS to be qualified for their duty position in both Army RC—the USAR and ARNG—at the start of FY95, along with the number who show a need for IET or reclassification (DMOSQ) training. As shown in the table, approximately 382,000 soldiers (about 78 percent of on-hand personnel) appeared to be qualified for their duty positions.[4] The percentage of DMOSQ personnel was slightly higher for the ARNG (80 percent) than for the USAR (75 percent), according to these data.

Table 3.1

MOS Reclassification Training Requirement by Component, FY95

Component	Number of Drilling Reservists	Number DMOSQ	Percent DMOSQ	Number Needing IET	Percent Needing IET	Number Needing Reclass	Percent Needing Reclass
ARNG	301,308	240,068	79.7	18,963	6.3	42,277	14.0
USAR	188,662	141,634	75.1	13,762	7.3	33,266	17.6
Total	489,970	381,702	77.9	32,725	6.7	75,543	15.4

SOURCE: ARNG and USAR SIDPERS, November 1995.

We define soldiers as needing IET if their duty MOS equals their primary/secondary/additional MOS and if their skill level is coded "0." We further examine these soldiers to confirm that they are non-prior-service personnel (using SIDPERS data for grade, pay entry base date, and time in active federal service) and have not already completed IET (according to ATRRS IET attendance records).

[3]Unlike the "select-train-promote" policy for NCOES, there is no clear policy identifying which soldiers should receive reclassification training before or following a job change and when this should occur. This makes it problematic to use a "stock and flow" approach to estimating training requirements, such as that used for NCO education in Chapter Two. However, when those who leave the service during a year are removed from the training requirement for that year, using successive annual snapshots yields estimates of training requirements similar to those produced by assuming that about 25 percent of the IET load is trained in the year that it first occurs and between 25 and 50 percent of the reclassification load is trained in the first year.

[4]These numbers are calculated using ARNG and USAR SIDPERS. As shown in Appendix A, these SIDPERS estimates are reasonably accurate for providing overall estimates of DMOSQ and non-DMOSQ soldiers across career management fields and MOSs.

The remaining soldiers show a need for training (approximately 108,000 or 22 percent), according to Army personnel records. Of these, 32,725 (6.7 percent) show as needing to complete IET, while the remaining 75,543 (15.4 percent) had completed IET and show a need for reclassification training.[5] This 15.4 percent, then, reflects the mismatch between the duty position these personnel are assigned to and any "earned" MOS. These soldiers require reclassification training, which in most cases would be provided at an RC training institution.[6]

As shown in Table 3.1, the number of soldiers who showed a need for reclassification training was sizable for both the ARNG and USAR, totaling as many as 75,543 enlisted personnel in FY95. Although the ARNG's requirement was numerically larger, exceeding the USAR's by about 9,000 trainees, the percentage of USAR soldiers who required reclassification was larger—17.6 percent, as compared to 14 percent at the start of FY95.

Qualification Levels Vary by Functional Area

To examine more closely the need for reclassification training, we disaggregated the DMOSQ training requirements shown in Table 3.1 according to functional area. Functional area is defined consistent with how schools are organized in the TASS, which aligns career management fields (CMFs) and MOSs with school brigades and battalions. A single brigade in each region of the country is responsible for conducting training in a designated functional area (e.g., for

[5]The "true" training requirement at any point in time may be different, given RC turbulence that occurs throughout the year. In general, attrition increases initial entry training requirements, while job turnover and non-prior-service accessions increases reclassification training requirements. In addition, some soldiers shown as needing training may have begun but not yet completed training, while others may attrit. These issues are discussed in more detail in Chapter Four.

[6]Most drilling reservists receive DMOSQ training in RC training institutions. However, some DMOSQ courses are unavailable at RCTIs because of the need for special equipment, funding limitations, lack of courseware, and so forth. In our baseline assessment, we noted DMOSQ courses were not available in RCTIs for approximately 15 percent of non-DMOSQ RC soldiers at that time. Hence, some of the 75,543 soldiers needing reclassification training in FY95 could not be trained in the RC because courses were unavailable at RCTIs. They may be available at Active Component schools.

combat arms CMFs), while battalions within the brigade are responsible for conducting individual training in specific branches (e.g., infantry, armor, and artillery). Other brigades are responsible for conducting training in the areas of combat support, combat service support, and health services, while additional brigades are responsible for officer education and NCO professional development courses.

Table 3.2 shows the number of drilling reservists who required reclassification training in each so-called TASS functional area. As the table shows, in sheer numbers, the largest reclassification training requirements were in the combat service support area (which is, in turn, composed of MOSs in the personnel support services, quartermaster, transportation, and ordnance branches).

CAPACITY IS NOT USED EFFICIENTLY IN MEETING DMOSQ TRAINING REQUIREMENTS

To assess the performance of the school system in meeting the DMOSQ training requirements, we compare these training requirements against the number of DMOSQ school quotas available in RC schools. Table 3.3 shows that when we compare the DMOSQ requirement of 75,543 against quota allocations, there was a significant shortfall—only 36,631 quota allocations were available at RCTIs in

Table 3.2

MOS Reclassification Training Requirement by Functional Area in FY95

Functional Area	Number Needing Reclassification	Percent Needing Reclassification
Combat arms	14,062	12.0
Combat support	18,187	16.3
Combat service support	35,148	17.4
Health services	5,784	12.2
Other[a]	2,362	19.5
Total	75,543	15.4

SOURCE: ARNG and SIDPERS, November 1995.

[a]The "other" category contains MOSs that fall outside traditional functional areas, such as recruiter and band member.

the nation as a whole (48 percent of the requirement at the start of the fiscal year). However, as was the case with NCOES training seats, a significant number of those quotas went unused during FY95. Overall, the number of student inputs was only 69 percent of the 36,631 quotas allocated (Table 3.3). Seventeen percent of the quotas (6,300) were unfilled, and an additional 14 percent (5,139) were lost because classes were cancelled. Altogether a little more than 25,000 quotas were used (i.e., had student inputs), accounting for 69 percent of the quotas allocated.[7]

The reasons that DMOSQ quotas went unused are similar to the reasons discussed in Chapter Two for NCOES courses. Our interviews and observations indicated that course cancellations often occurred because of shortages of qualified instructors or failures to achieve the

Table 3.3

DMOSQ Quota Allocations by Functional Area in FY95

Course	Number Needing Reclass	Quota Allocations *(Percent of Needing Reclass)*	Quotas Cancelled *(Percent of Allocation)*	Quotas Unfilled *(Percent of Allocation)*	Total Quotas Used (Inputs) *(Percent of Allocation)*
Combat arms	14,062	7,534 *(54)*	754 *(10)*	1,649 *(22)*	5,131 *(68)*
Combat support	18,187	9,015 *(50)*	919 *(10)*	1,289 *(14)*	6,807 *(76)*
Combat service support	35,148	18,297 *(52)*	3,389 *(19)*	2,971 *(16)*	11,937 *(65)*
Health services	5,784	972 *(17)*	30 *(3)*	179 *(18)*	763 *(79)*
Other	2,362	813 *(34)*	47 *(6)*	212 *(26)*	554 *(68)*
Total	75,543	36,631 *(48)*	5,139 *(14)*	6,300 *(17)*	25,192 *(69)*

SOURCE: ATTRS School Aggregate file November 1995.

[7]In the timeframe analyzed, ATRRS records of reclassification training appear to miss some soldiers who take required classes as walk-ons. Thus, ATRRS records may understate somewhat the full level of quota utilization.

minimum enrollment needed to conduct the class. This latter problem can result from inaccurate forecasts of training requirements for specific MOSs, as well as from problems in using the reservation system (ATRRS). Unfilled seats occurred because of lack of experience in using the reservation system, "no-shows" (i.e., soldiers who make reservations and do not show up for school), and misallocations of quotas to units; these misallocations involve "overallots" to some and "underallots" to others, which could have prevented qualified soldiers from attending class.

In terms of production of graduates in FY95, we saw further evidence of underutilized capacity. As shown in Table 3.4, the graduation rate for the 25,192 inputs was 94 percent, which was fairly consistent across functional areas, but this represents only 65 percent of the initial quotas overall. Production varied considerably across different functional areas, given differences in quota fill and graduation rates. Ultimately, the 23,758 graduates represents only 31 percent of the number of soldiers shown in Table 3.3 as needing reclassification training.

SCHOOL CAPACITY TO CONDUCT DMOSQ TRAINING IMPROVED ACROSS FISCAL YEARS

When we turn to the question of how well school capacity matched the training requirement (as above, using the 75,543 training requirement at the start of FY95), we see that despite the continuing size of the DMOSQ requirement and continuing problems with the use of training capacity, the problem diminished from FY94 to FY95. Figure 3.1 shows that on the delivery side, the number of quota allocations for DMOSQ courses increased considerably, from 31,619 to 36,631—an increase of 16 percent. Meanwhile, the number of non-DMOSQ soldiers fell from the start of FY94 to the start of FY95 (from 87,985 to 75,543), helping to close the gap between the requirement and capacity of schools to meet that requirement.[8] Specifically, in

---

[8]The number of drilling reservists fell by 4.7 percent, from 514,025 in FY94 to 489,970 in FY95. At the same time, the number of non-DMOSQ soldiers fell disproportionately more—by 14.1 percent from the start of FY94 to the start of FY95 (from 87,985 to 75,543). This, however, proved to be a transitory phenomenon, as will be discussed later.

Table 3.4

DMOSQ Production by Functional Area in FY95

Functional Area	Quota Allocations	Inputs	Graduates	Graduation Rate	Ratio of Grads to Quotas
Combat arms	7,534	5,131	4,682	.91	.62
Combat support	9,015	6,807	6,530	.96	.72
Combat service support	18,297	11,937	11,273	.94	.62
Health services	972	763	724	.95	.74
Other	813	554	549	.99	.68
Total	36,631	25,192	23,758	.94	.65

SOURCE: ATTRS School Aggregate file November 1995.

FY95, quota allocations in MOS-producing courses met 48 percent of the requirement versus 36 percent of the requirement in FY94.

Problems Remained with Capacity Utilization

Although capacity in relation to requirements improved across the fiscal years examined in the study, serious problems remained with the use of school quotas to meet DMOSQ training requirements. As shown in Figure 3.2, a little less than one-third of the initial DMOSQ quotas were lost during FY94, attributable in pretty much equal parts to cancelled courses and to unfilled seats in the classes that were held. Quota use was very slightly improved but essentially the same in FY95. The overall system wasted much of its available capacity, which again reflects insufficient use of the reservation system and insufficient emphasis on sending non-DMOSQ soldiers to school.

The Trend Varied by Functional Area

The alignment of DMOSQ training supply and demand differs by functional area, and consistent with the overall figures, tended to be higher in FY95, as compared to FY94. Specifically, the number of quota allocations in combat arms rose from 38 percent of the requirement in FY94 to 54 percent; combat support, from 37 to 50; and

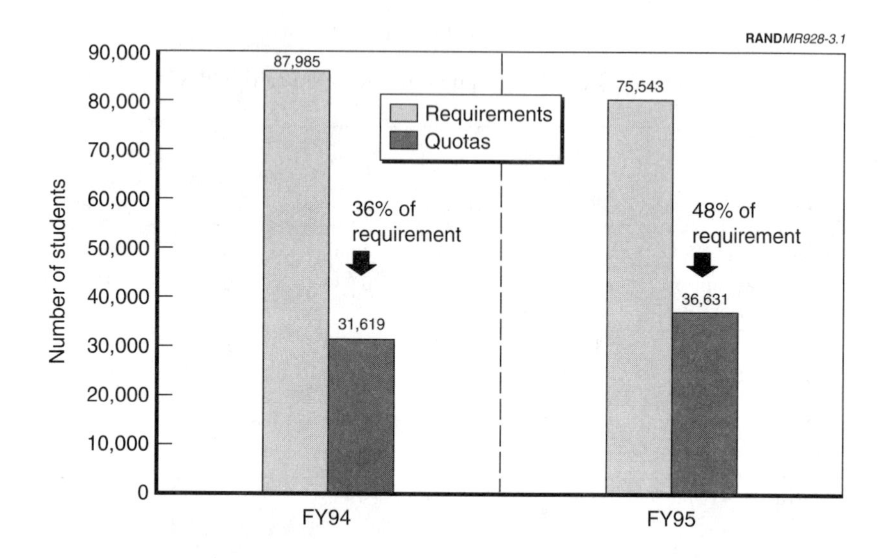

Figure 3.1—In DMOSQ Courses, Demand Is Down and Capacity Is Up Across the Fiscal Years

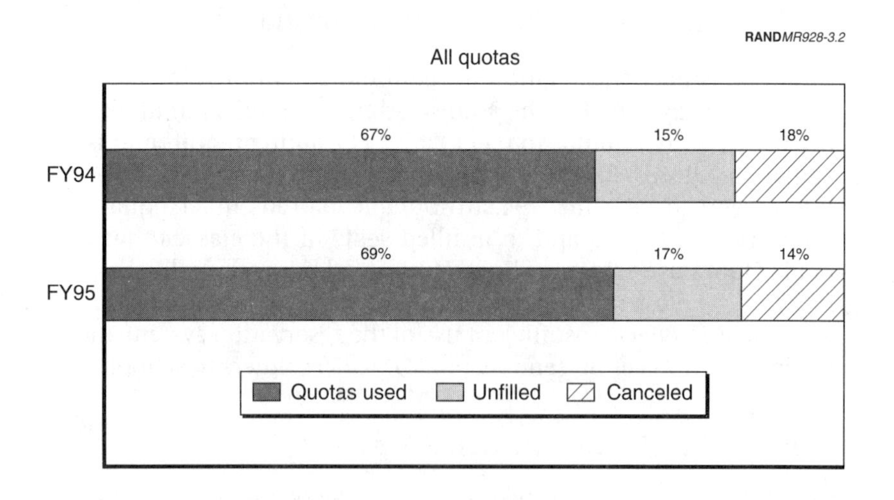

Figure 3.2—Quota Use Remained Inefficient Across the Fiscal Years

combat service support, from 37 to 52. The only exception is health services, where the number of quotas dropped from 20 percent of the requirement in FY94 to 17 percent in FY95.[9]

In terms of cancelled quotas, combat arms showed improvement, falling from 16 percent to 10 percent; it was followed by combat support (14 to 10). Health services (5 to 3) and combat service support (22 to 19) remained about the same, with the latter showing the highest percentage of cancelled quotas among the various functional areas.

In terms of unfilled quotas, combat support and combat service support remained about the same in FY95 as compared to FY94 (14 versus 13 percent; 15 versus 16 percent, respectively), and thus tracked the national trend. However, combat arms actually got worse in this regard, going from 19 percent to 22 percent, while health services got better, falling from 24 percent to 18 percent.

MOSQ Production Rates Remained Constant Across Fiscal Years

Figure 3.3 compares the production of graduates from DMOSQ courses across fiscal years. As Figure 3.1 showed, the school system did significantly better in terms of quota allocations versus requirements in FY95, because requirements fell and quota allocations rose; as a result, in FY95, the allocations represented 48 percent of the overall DMOSQ training requirement versus only 36 percent in FY94. As Figure 3.2 showed, the fill rate for quota allocations was roughly the same in FY95 as in FY94 (69 versus 67 percent). Consequently, as shown in Figure 3.3, given an equivalent graduation rate, the school system used its quotas at about the same rate to produce graduates across fiscal years. In FY95, 65 percent of the quotas were used to produce graduates (23,758 graduates from 36,631 quotas) versus 63 percent in FY94 (19,933 graduates from 31,691 quotas).

Although quota fill rates were equivalent, because quotas and requirements were in better alignment, the system produced more

[9]Again we note that requirements fell to a greater extent than the overall force was reduced during this period.

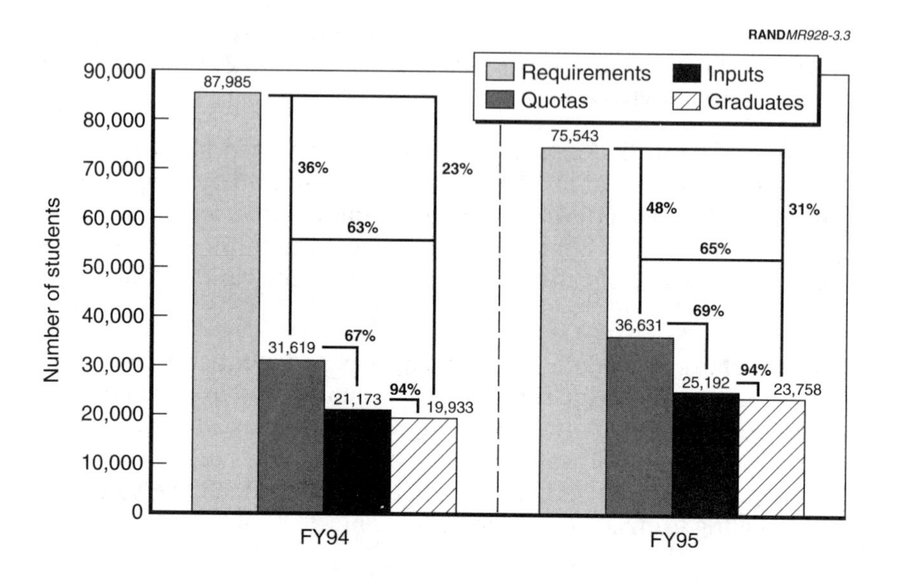

Figure 3.3—The System Did Better in Producing DMOSQ Graduates Across the Fiscal Years

graduates in relation to the DMOSQ training requirement in FY95. Still, when all is said and done, the output of the school system in FY95 still amounts to only 31 percent of the total estimated reclassification training requirement (23,758 graduates versus a requirement of 75,543), compared to 23 percent in FY94 (19,933 graduates versus a requirement of 87,985).

REGION C COMPARES WELL TO THE REST OF THE NATION

When we compare the DMOSQ rate in Region C to other regions of the nation in FY95, we see similar levels of duty MOS qualification: 79.2 percent versus 77.7 percent, respectively. In terms of the number of soldiers who appear to need reclassification training, Table 3.5 compares Region C with the rest of the nation (excluding Region C). The percentage of Region C soldiers showing a need for reclassification training, compared to other regions of the nation, is fairly similar (14.4 percent versus 15.6 percent, respectively).

Table 3.5 also shows how these soldiers are distributed across the various functional areas. Region C soldiers need reclassification training to a slightly lesser extent in the combat support, combat service support, and health services MOSs. They are equally qualified in combat arms and less qualified in "other" functional areas.[10]

Table 3.6 shows how well quotas were used in Region C in FY95, overall and in each functional area. As the table shows, 81 percent of the quotas allocated in Region C were filled by students; this was much higher than in other regions of the nation (where the quota fill rate was 68 percent). This seems attributable to a higher fill rate in Region C in combat arms courses (97 percent in Region C, compared to 65 percent elsewhere) and combat service support courses (76 percent in Region C, compared to 64 percent elsewhere). While the percent of cancelled quotas in Region C was comparable with other regions (16 percent versus 14 percent), the percentage of unfilled quotas was substantially smaller (3 percent versus 19 percent).

Table 3.7 shows the graduation rate for student inputs in Region C, along with the ratios of graduates to initial quota allocations. As

Table 3.5

MOS Reclassification Training Requirement by Functional Area in FY95: Region C Versus Other Regions

	Other Regions		Region C	
Functional Area	Number Needing Reclass	Percent Needing Reclass	Number Needing Reclass	Percent Need Reclass
Combat arms	12,446	11.9	1,616	12.7
Combat support	16,292	16.5	1,895	14.7
Combat service support	31,457	17.7	3,691	15.2
Health services	5,269	12.5	515	9.9
Other	1,909	18.4	455	26.9
Total	67,373	15.6	8,172	14.4

SOURCE: ARNG and SIDPERS, November 1995.

[10]As mentioned earlier, the "other" category contains MOSs that fall outside traditional functional areas, such as recruiter and band member.

Table 3.6

DMOSQ Quota Allocations by Functional Area in Region C in FY95

Functional Area	Number Needing Reclass	Quota Allocations *(Percent of Number Eligible)*	Quotas Cancelled *(Percent of Allocation)*	Quotas Unfilled *(Percent of Allocation)*	Total Quotas Used *(Percent of Allocation)*
Combat arms	1,616	694	94	0	672
		(43)	*(14)*	*(0)*	*(97)*
Combat support	1,895	556	56	66	434
		(29)	*(10)*	*(12)*	*(78)*
Combat service support	3,691	1,903	372	76	1,455
		(52)	*(20)*	*(4)*	*(76)*
Health services	515	52	1	11	40
		(10)	*(6)*	*(17)*	*(77)*
Other	455	155	0	36	119
		(34)	*(0)*	*(23)*	*(77)*
Total	8,172	3,360	523	117	2,720
		(41)	*(16)*	*(3)*	*(81)*

SOURCE: ATTRS School Aggregate file November 1995.

shown in the table, 88 percent of students enrolled in courses graduated in Region C—a rate that is lower than the other regions (where the graduation rate was 95 percent). However, given the higher fill rates in Region C (81 percent, as shown in Table 3.6, versus 68 percent in other regions), the overall output of the school system was higher in Region C than elsewhere. Overall, graduates in Region C represented 71 percent of quota allocations, versus 64 percent in other regions.

The foregoing discussion highlights school system performance during FY95. We also examined the time trends in Region C schools as compared to the rest of the nation. We first observe that the overall DMOSQ rate rose from 75.9 percent to 79.2 percent in Region C from the beginning of FY94 to the start of FY95 (compared with 75.2 to 77.7 for other regions). At the same time, the proportion of soldiers showing a need for reclassification training fell from 16.9 percent to 14.4 percent (versus a change from 17.2 to 15.6). This trend is consistent across the functional areas, although levels vary.

Table 3.7

DMOSQ Production by Functional Area in Region C in FY95

Functional Area	Quota Allocations	Inputs	Graduates	Graduation Rate	Ratio of Grads to Quotas
Combat arms	694	672	443	.66	.64
Combat support	556	434	427	.98	.77
Combats ervice support	1,903	1,455	1,367	.94	.72
Health services	52	40	36	.90	.69
Other	155	119	119	1.0	.77
Total	3,360	2,720	2,392	.88	.71

SOURCE: ATTRS School Aggregate file November 1995.

Quota use in Region C was very good compared to the other regions of the nation. It improved on the trend elsewhere in the nation for cancelled quotas (falling from 22 percent to 16 percent, compared to a decrease from 17 to 14 percent in other regions). The proportion of unfilled quotas increased to an extent similar to other regions, rising from 2 percent to 3 percent (compared to a rise from 16.5 to 19.5 percent in other regions). Functional area trends in Region C were consistent with national trends, except that reductions in cancelled quotas was significantly greater in combat arms (26 to 14) and combat support (32 to 10) courses than they were in other regions.

Finally, when we look at Region C production compared to the rest of the nation, we see that it did about the same. Specifically, graduates represented 71 percent of quota allocations in FY95 (up from 69 percent in FY94).

DMOSQ TRAINING REQUIREMENTS AT THE END OF FY95

The results presented in this chapter suggest improvements in schools' ability to meet DMOSQ requirements, as well as equivalent or better performance by schools conducting reclassification training in Region C and elsewhere in FY95 (compared to FY94). However, an interesting development occurred with respect to DMOSQ rates at the beginning of FY96. While requirements for DMOSQ training fell from the start of FY94 to the start of FY95 (the period of observation), by the start of FY96, they had actually gotten worse again. This is shown in Figure 3.4, which compares the number of DMOSQ and non-DMOSQ soldiers at the start of FY95 and the end of FY95 (i.e., the start of FY96). The numbers and percentages of soldiers showing need for reclassification training actually rose. The end result was that the DMOSQ rate itself fell from 77.9 percent to 75.3 percent during the fiscal year.[11]

How did it happen that the DMOSQ rates fell and reclassification training requirements rose in the face of a school system whose performance remained constant or improved relative to the previous

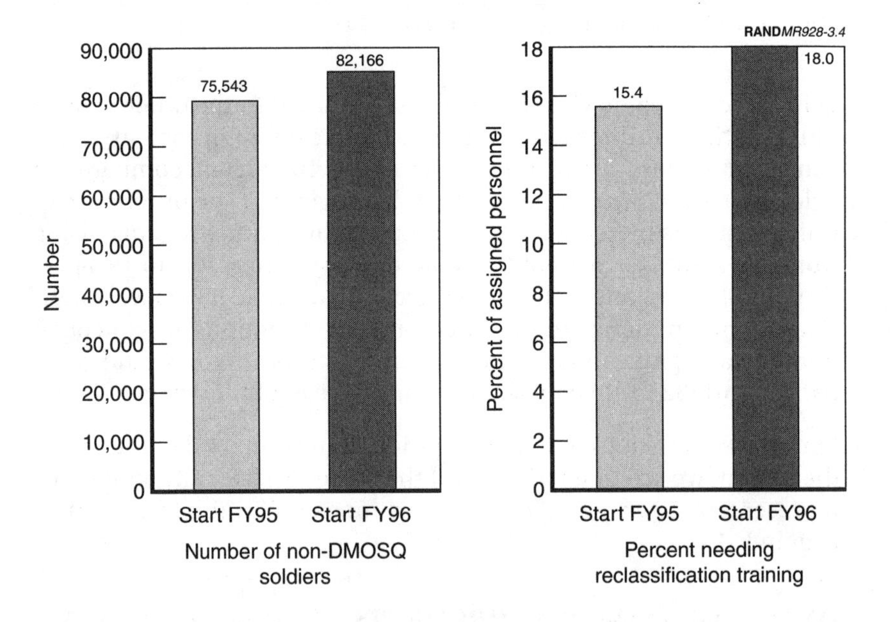

Figure 3.4—The Number of Non-DMOSQ Soldiers Rose in Number and Percentage by the Start of FY96

---

[11]The proportion of assigned personnel who are non-DMOSQ rises considerably (from 15.4 to 18.0 percent) because end strength fell as the number of non-DMOSQ personnel rose. These personnel dynamics are discussed in greater detail in Chapter Four.

fiscal year? The reason for the fall-off in DMOSQ rates and the increase in reclassification training requirements stems from personnel turbulence that is endemic to the RC and ultimately speaks to potential steps for managing training requirements more effectively.

Data from SIDPERS show that during the course of FY95, personnel turbulence set back the training system in the following ways: during the year, as end strength fell by about 34,000 (including qualified soldiers and some non-DMOSQ soldiers who left the force), about 14,600 new prior-service soldiers appear in the personnel records in positions for which they were not DMOSQ.[12] Other soldiers remained in the system but changed jobs—29,000 soldiers changed their duty MOS away from positions for which they had been qualified to ones for which they are not qualified.[13] Meanwhile, RC schools turned out 24,000 reclassification graduates. The net effect of personnel movements, attrition, and training raised DMOSQ training requirements and lowered DMOSQ, as shown in Figure 3.4.

These findings indicate that changing the organization and management of RC schools, as embodied in the new structures of the TASS, will not on its own decrease training requirements and improve readiness. As these are appropriate goals for a school system to attain, additional measures should be taken to reduce demands on the system. The next chapter will discuss the problem of personnel turbulence in greater detail. There we will also address the potential benefits of incorporating measures to reduce personnel turbulence as a way to help the Army school system become more efficient and effective in meeting training requirements.

IMPLICATIONS

The findings presented in this chapter indicate that, like NCO training, some improvements occurred but deep-seated structural problems remained with the management of RC DMOSQ training requirements and with the use of school capacity to meet these requirements. Again, there was a major problem both with the size

---

[12]Additional non-prior-service soldiers took positions for which they were MOS qualified.

[13]Also, some non-DMOSQ soldiers moved to positions for which they were qualified.

of the requirement (especially in relation to supply of training) and with the efficient use of available training capacity to meet these requirements. In particular, even as training requirements remained large, we saw continuing problems with the use of school quotas that are available to meet these requirements.

In addition, however, DMOSQ training faces a particular systemic problem with respect to personnel turbulence and attrition, which drives up training requirements and makes it difficult (if not impossible) for the school system to make headway against its requirements. Hence, policies that improve the management of personnel must be integrated with those that govern school operations to reach Army objectives for readiness and to ensure a smoothly operating school system to support this. We now turn to some of the benefits of such an approach.

Chapter Four

HOW REDUCING TURBULENCE AFFECTS THE DMOSQ TRAINING REQUIREMENT

As Chapter Three indicated, reclassification (DMOSQ) training faces both demand-side and supply-side problems. On the demand side, there is a significant requirement each year for soldiers needing DMOSQ training (75,543 at the start of FY95); on the supply side, there are too few quota allocations to meet the demand (36,631 at the start of FY95, or only 48 percent in relation to the estimated requirement), and even fewer of those allocations are actually used to produce graduates (23,758 in FY95, or 31 percent in relation to requirements). The dual nature of the problem suggests that there are two different avenues available for solving it. One can attack the supply side of the problem and try to improve the delivery of training and production of graduates—a current thrust of the TASS. In our recommendations (Chapter Five) we address ways to address this supply problem, including developing new requirements estimation and forecasting tools and increasing oversight of training execution.

One can also attack this problem from the demand side and search for ways to reduce training requirements. For example, according to an ORE audit conducted as part of the effort to validate our SIDPERS-based estimates of training requirements (see Appendix A), only a quarter of non-DMOSQ soldiers in FY95 were actively in the training pipeline—scheduled, enrolled, on wait-lists, etc. About 50 percent of non-DMOSQ soldiers showed no indication that they were being programmed for training, and the other 25 percent were in this state because of "personnel churn"—there was uncertainty about whether they were staying or leaving or changing jobs inside the unit, or moving to another unit. Although such personnel churn—which causes attrition and turbulence—creates problems for the RC schools that changes to school organization, management, and oversight cannot solve on their own, personnel churn can still be tackled, which, in turn, can reduce the requirement.[1]

Indeed, problems of turnover in the RC personnel system are of long-standing concern and have been examined in several previous research studies.[2] Prior research shows that personnel turnover is a chronic problem in the RC, occurring both when force structure is changing and when it is relatively stable. This research also shows that attrition is an important factor that lowers DMOSQ rates in RC units and increases the need for training. The research also examines job movements in the RC and shows that most job movements occur within units and within the local area, rather than from long geographic moves and resulting changes in unit affiliation. Finally, previous research has analyzed the impact of reducing job turnover and attrition on unit readiness and training requirements, demonstrating significant benefits from tackling these personnel problems.[3]

This chapter builds on this research and takes another look at the degree to which controlling the personnel churn associated with attrition and job movements in the RC can reduce DMOSQ requirements, in light of the TASS reorganization that is under way. The analyses examine the degree to which turnover is personnel-driven or is caused by changes in force structure. After defining several states that characterize the status of RC soldiers, we examine personnel movements into, out of, and within the RC personnel system based on two years of SIDPERS data (from the beginning of FY94 to the end of FY95) to get at the issue of what causes the personnel churn: is it due to force structure changes that require currently qualified soldiers to take new jobs for which they are not qualified, or

---

[1]The Army's "select-train-promote" policy is another way of reducing training requirements, since this policy limits NCO education to only soldiers who are selected for promotion.

[2]Richard J. Buddin and David W. Grissmer, *Skill Qualification and Turbulence in the Army National Guard and Army Reserve*, Santa Monica, CA: RAND, MR-289-RA, 1994; Ronald E. Sortor, Thomas F. Lippiatt, J. Michael Polich, and James C. Crowley, *Training Readiness in the Army Reserve Components*, Santa Monica, CA: RAND, MR-474-A, 1994.

[3]Bruce R. Orvis et al., *Ensuring Personnel Readiness in the Army Reserve Components*, Santa Monica, CA: RAND, MR-659-A, 1996.

is it due to personnel movements from one job to another in stable units?

We then examine how the personnel flows differ for units that are being activated, inactivated, or converted in some way versus units that are "stable" (i.e., not undergoing a unit modification of some type). Based on the personnel flows during the two years, we then derive transition probabilities and use them in a stochastic Markov model to predict the future steady-state populations in each of the states and, therefore, the number of soldiers that will require reclassification training. Finally, we estimate the impact on the demand for training of actions designed to reduce personnel attrition or unit turbulence.

DEFINING SYSTEM STATES

In Chapter Three we used a "snapshot" of training requirements at the start of the fiscal year and compared them to school capacity and output during that year. In fact, estimating training requirements with precision is far more complex, as the number of personnel needing training changes continually as soldiers enter and leave the force or move from job to job throughout the year. Also, soldiers shown in personnel records as needing training may be at different points in the training pipeline—some having not yet started, others in process.

To model these phenomena, however, we need to make some simplifying assumptions. First, we assume that SIDPERS records in the USAR and the ARNG are reasonably accurate in portraying a soldier's duty MOS and primary MOS, as these are compared to determine DMOSQ and training status.[4] Second, we assume that if we examine "snapshots" of the force at different points in time, we can characterize with reasonable accuracy the changes in force composition and DMOSQ and training status that occur.[5]

---

[4]We conducted an audit of these records during ORE visits to selected units and found them highly accurate. See Appendix A for details.

[5]Our transition statistics are based on "snapshots" and don't take into account other, potentially relevant issues. These include assumptions as to how quickly and with what priority any "carryover" requirement will be trained and the rate at which the USAR or ARNG aim to complete reclassification and initial entry training. Improve-

To characterize the relationship between personnel turbulence and training requirements, we define four potential states to describe the movement of RC personnel from one time period to the next. In the first state, a soldier can be "not present" in either the initial or the subsequent time period. This state captures soldiers who enter or leave the system. For example, a soldier who was not present (i.e., does not appear in the SIDPERS file) in one time period but is an RC member in the subsequent time period (i.e., does appear in the following year's SIDPERS file) represents an addition to the personnel inventory. On the other hand, soldiers who were present in one time period but not in the next represent a loss to the system.

The other three states are duty MOS qualified (DMOSQ), "Need IET," and non–duty MOS qualified (NDMOSQ). Personnel in the need IET category represent a demand for initial skill training that is currently satisfied in Active Component (AC) training institutions, while personnel in the NDMOSQ state represent an immediate demand for reclassification training that can, in most cases, be satisfied in an RC school.[6] To further understand personnel movements within the RC system, we break the two states (NDMOSQ and DMOSQ) into two substates indicating whether the soldier's duty MOS stayed the same (same MOS) or changed (changed MOS) from one time period to the next.[7]

PERSONNEL MOVEMENTS FOR FY94 AND FY95

Using these four states (and two substates), we can match the status of a person in one time period with his status in the subsequent time period to provide a measure of the flow of soldiers between the dif-

ments are needed in the methods used to calculate training requirements; such improvements could provide more precise estimates given additional assumptions.

[6]As indicated earlier, RC schools currently provide most but not all courses needed to reclassify RC soldiers in a new MOS.

[7]Soldiers who were not present in the initial year but enter the RC as DMOSQ or non-DMOSQ in the subsequent year are shown as "same MOS" in these tables. In addition, readers will note some discrepancies in the data shown in Tables 4.2 and 4.3 (e.g., some soldiers are shown as DMOSQ in one year and as needing IET the next; others are shown as DMOSQ in the initial year and non-DMOSQ in the same MOS the subsequent year). These appear to be data errors and are not corrected in the tables. They affect a very small number of records and do not affect the calculations shown later in this chapter.

ferent states. Table 4.1 shows the results for FY94 across these states (from the start of FY94 to the start of FY95), for all drilling reservists serving in RC units.[8] For example, reading across the first row of the table, 82,411 new soldiers joined RC units as drilling reservists in FY94; reading down the first column, 106,466 drilling reservists shown in personnel records at the start of the fiscal year left the force during FY94.[9] The starting force size for a given year is the difference between the total number of soldiers tracked over the one-year period and the number of soldiers who entered during the time period. In FY94 that number is 514,025, or 596,436 minus 82,411. The DMOSQ rate at the beginning of a year is the number of soldiers who were DMOSQ at the beginning of the year divided by the force size at the beginning of the year. For FY94 that rate is 75.3 percent, or 387,042 divided by 514,025.

Table 4.2 shows the same information for FY95 (from the start of FY95 to the start of FY96). The starting force size is 489,970 (565,565 minus 75,595) and the DMOSQ rate is 77.9 percent (381,702 divided by 489,970).

Table 4.1

Personnel Movements Between States (FY94)

	Start of FY95						
			NDMOSQ		DMOSQ		
Start of FY94	Not Present	Need IET	Same MOS	Changed MOS	Same MOS	Changed MOS	Total
Not present	0	24,819	12,787	0	44,805	0	82,411
Need IET	10,415	3,974	646	1,522	20,576	1,865	38,998
NDMOSQ	20,208	1,019	26,599	9,450	15,389	15,320	87,985
DMOSQ	75,843	2,913	996	23,543	265,900	17,847	387,042
Total	106,466	32,725	41,028	34,515	346,670	35,032	596,436

[8]The time periods shown are from the start of FY94 to the start of FY95, hence the table captures changes that occur during FY94.

[9]Hence losses exceeded gains by approximately 24,000 during FY94, reflecting that this period was one in which force structure was being reduced.

Table 4.2

Personnel Movements Between States (FY95)

	Start of FY96						
			NDMOSQ		DMOSQ		
Start of FY95	Not Present	Need IET	Same MOS	Changed MOS	Same MOS	Changed MOS	Total
Not present	0	22,551	14,619	0	38,425	0	75,595
Need IET	10,191	4,483	593	1,449	14,860	1,149	32,725
NDMOSQ	18,929	492	26,684	8,162	11,544	9,732	75,543
DMOSQ	80,300	2,819	1,714	28,945	250,386	17,538	381,702
Total	109,420	30,345	43,610	38,556	315,215	28,419	565,565

The DMOSQ rates for the time periods represented in the data (from the start of FY94 to start of FY96) are shown in Table 4.3. The DMOSQ rates for FY94 and FY95 are calculated as described above. We also show the DMOSQ rate at the start of FY96, which is calculated from Table 4.2 based on a starting force size of 456,145 (565,565 minus 109,420) and the number of MOS-qualified soldiers at the start of FY96 (343,634, the sum of 315,215 and 28,419). The DMOSQ rate is 75.3 percent (343,634 divided by 456,145).

Our interest in this chapter centers on the movement of soldiers into and out of the DMOSQ states described earlier—movements that affect the proportion of the RC force that is duty qualified at any point in time. In particular, we are interested in changes in the number of

Table 4.3

DMOSQ Rates (FY94–FY96)

Start of Fiscal Year	DMOSQ Rate
1994	75.3%
1995	77.9%
1996	75.3%

DMOSQ soldiers (increases and decreases), as well as overall attrition, attrition of DMOSQ soldiers, the number of soldiers moving from the state of not present to the state of DMOSQ (qualified entrants), the number of soldiers moving from the NDMOSQ and need IET (nonqualified) states to the DMOSQ state, and the number of soldiers moving from the DMOSQ state to the nonqualified states. These are of interest in their own right and as the basis for modeling the impact of reducing personnel movements, as will be described later in this chapter.

For example, during FY94, we see from looking at Table 4.1 that DMOSQ increases and decreases almost balance out. In terms of increases, adding up all those in the not present, need IET, and NDMOSQ rows that became DMOSQ either in the same MOS or a new MOS (44,805, 20,576, 15,389, 1,865, 15,320) yields 97,955 DMOSQ soldiers. DMOSQ decreases—the total of the DMOSQ row minus those soldiers who remained DMOSQ in either the same or a new MOS (387,042 minus 17,847 minus 265,900)—amount to 103,295 soldiers.

In terms of overall attrition during FY94, dividing the total not present at the end of the year (106,466) by the force size at the beginning of the year (514,025) yields an attrition rate of 20.7 percent. DMOSQ attrition, which is calculated by dividing the total DMOSQ (387,042) into DMOSQ in the not present state (75,843)—is 19.6 percent. Thus, attrition rates for DMOSQ soldiers were about the same as the overall attrition rate.

Table 4.1 also provides the basis for determining other personnel dynamics of interest. For example, the percentage of new accessions who enter the RC as duty MOS qualified (or are fully trained during the year of entry) is determined by dividing 44,805 by 82,411, which equals 54.4 percent. The percentage of NDMOSQ/need IET soldiers who became DMOSQ—the ratio of those NDMOSQ and need IET who became DMOSQ in either the same or a new MOS (53,150) to the total for the NDMOSQ/need IET rows (126,983)—is 41.9 percent. Finally, the percentage of DMOSQ soldiers who stay in the force but no longer show as MOS qualified—the ratio of those in the DMOSQ row who need IET or who are NDMOSQ in either the same or a new

MOS (27,452) to the total number of DMOSQ (387,042)—is 7.1 percent.[10]

These personnel movements and other relevant transition percentages for FY94 and FY95 are presented in Table 4.4.

The data show that in both time periods, the RC lost more duty qualified soldiers then they gained. During FY94, DMOSQ decreases were only slightly higher than increases, and since the RC end strength was declining, the overall DMOSQ rate for the force actually increased by the start of FY95 (from 75.3 percent at the start of FY94 to 77.9 percent at the start of FY95). However, the DMOSQ rate dropped during FY95—from 77.9 percent at the start of the year to 75.3 percent by the start of FY96. DMOSQ decreases were substantially larger than increases, fewer duty qualified prior-service soldiers (those not present in a time period) were recruited, and a smaller percent of the force became duty qualified during the year.

Table 4.4

Transition Statistics for FY94 and FY95

Transition Statistics	FY94	FY95
DMOSQ increases	97,955	75,710
DMOSQ decreases	103,295	113,778
Overall attrition (%)	20.71	22.33
DMOSQ attrition (%)	19.60	21.04
Not present to DMOSQ (%)	54.37	50.83
Not qualified to DMOSQ (%)	41.86	34.44
DMOSQ to not qualified (%)	7.09	8.77

[10]As reflected in Table 4.1, Army personnel records show 2,913 previously DMOSQ soldiers as needing IET and 996 previously DMOSQ soldiers as no longer DMOSQ in the same MOS in FY94. These are likely due to errors in the personnel records. However, we did not wish to treat these soldiers as qualified and hence categorized them as nonqualified, together with the 23,543 soldiers who changed MOS and became unqualified.

DETERMINING SOURCES OF PERSONNEL TURBULENCE

The time period we are examining was one of change in the RC force structure. Overall, end strength dropped over 10 percent and the types of units in the ARNG and the USAR were altered through unit inactivations, activations, and conversions. This turmoil within the RC structure may have had an impact on the movement of soldiers between the various states. Thus, the real question is, "Is the turbulence worsened by changes in the force structure or is it mostly personnel-driven?" In other words, was the period studied (1994–1995) simply an especially turbulent time as a result of force structure changes, or was the turbulence caused by chronic personnel factors endemic to the RC that affect individual soldiers' decisions to change jobs or units or leave the military (e.g., "better" job assignment, promotion opportunities, conflicts with civilian employment, household moves, or family pressures)?

The answer has to do with how to address the problem. If the problem is largely caused by personal decisions, then changes to personnel policies might remedy the problem; if the problem is that force structure changes significantly exacerbate the turbulence, then changes to personnel policy would be less useful. Instead, actions to reduce force structure turbulence, or at least defer intervention until force structure changes have completed their course, could prove productive.

Unit Status

To get at the source of the turbulence, we used Army force structure data (the MFORCE) to capture the status of a unit during each year of observation. We created four categories of units:

1. Stable—units that were not coded for activation, inactivation, or conversion during a year;
2. Inactivating—units that were present during the first year but not present during the next year;
3. Activating—units that were not present during the first year but were present during the next year;

4. Converting—units that were coded for conversions during the year.

Tables 4.5 and 4.6 capture the transition statistics by the four unit statuses defined above for the two fiscal year periods (1994 and 1995), respectively. The data associate the soldiers shown in Tables 4.1 and 4.2 with units shown in MFORCE as falling in the four states defined above. For example, Table 4.5 shows how the 514,025 soldiers who were present at the start of FY94 (596,436 minus 82,411 in Table 4.1) were distributed according to unit status (stable, inactivating, activating, converting). Table 4.6 shows similar information for the 489,970 soldiers who were present at the start of FY95 (565,565 minus 75,595 in Table 4.2).

Several observations can be drawn from the data in Tables 4.5 and 4.6. First, most of the soldiers were in stable units in both FY94 and FY95. Fully 84 percent of the soldiers in FY94 and 74 percent of the soldiers in FY95 were in units designated as stable. Therefore, the transition statistics for the force mirror the values for the stable units.

Table 4.5

Transition Statistics by Unit Status (FY94)

Transition Statistics	Total	Stable	Inactivating	Activating	Converting
Number of soldiers	514,025	429,429	5,496	1,685	77,415
Overall attrition (%)	20.71	20.69	24.31	—	21.04
DMOSQ attrition (%)	19.60	19.54	24.13	—	20.08
Not present to DMOSQ (%)	54.37	54.33	—	62.12	—
Not qualified to DMOSQ (%)	41.86	41.58	29.80	54.06	43.89
DMOSQ to not qualified (%)	7.09	6.67	18.25	16.81	8.58

NOTE: The number of soldiers is measured at the start of the year.

Table 4.6

Transition Statistics by Unit Status (FY95)

Transition Statistics	Total	Stable	Inactivating	Activating	Converting
Number of soldiers	489,970	364,255	26,040	5,784	93,891
Overall attrition (%)	23.99	22.26	27.69	—	22.51
DMOSQ attrition (%)	21.04	20.93	26.67	—	21.08
Not present to DMOSQ (%)	50.83	51.03	—	44.54	—
Not qualified to DMOSQ (%)	34.44	33.72	32.43	49.82	36.54
DMOSQ to not qualified (%)	8.77	7.50	22.34	26.21	8.95

NOTE: The number of soldiers is measured at the start of the year.

Second, only a small number of soldiers were with units that were activating or inactivating in either year. The various transition statistics appear logical when compared to the measures for the stable units. For example, the attrition-oriented measures are all higher for the inactivating units, suggesting a larger percentage of soldiers in inactivating units are leaving the force. For activating units, a high percentage of the soldiers are either moving into positions where they have the necessary qualifications or for which they are currently unqualified and will require reclassification training. Also, the attrition rate for activating units is lower than the rate for other types of units. Ultimately, even though the statistics for inactivating and activating units follow the expected pattern, their impact is minimal, because the two categories represent such a small portion of the soldiers involved (1 percent and 6 percent, respectively, in FY94 and FY95). Finally, the transition statistics for units undergoing conversions of some type are very similar to those for stable units.

MODELING HOW REDUCING PERSONNEL MOVEMENTS AFFECT DMOSQ TRAINING DEMAND

As mentioned earlier, if personnel movements, as opposed to changes in force structure, drive turbulence and attrition, then the movements we see over the two-year period provide a basis for estimating the potential impact on training requirements of reducing personnel movements. If we assume that the overall levels of personnel movement are roughly consistent with historical levels, then we can use a simple inventory projection model to estimate the future demand for duty-related training that would follow from a change in the rate of turbulence and attrition.[11]

Personnel flows during the course of a time period provide a measure of the transition probabilities between the various states for that period. These probabilities represent the likelihood of a soldier being in a particular state at the end of the year given having been in a specific state at the beginning of the year. They are calculated by dividing each cell entry shown earlier in Tables 4.1 and 4.2 by the row total for the specific cell. For example, during FY94 (Table 4.1), the probability that a soldier who started the year as DMOSQ would leave the force is 0.1960 (or 75,843 divided by 387,042). The transition probabilities for FY94 and for FY95 are shown in Table 4.7 and Table 4.8, respectively.

Given the number of soldiers in the various states at the start of a time period, we can use the transition probabilities to calculate the number in the various states at the end of the time period. If we assume the transition probabilities are constant from period to period, we can analytically calculate (through a Markovian process) the

---

[11]Because some downsizing occurred during this period, it is possible that the rates of job attrition and turbulence are somewhat higher than normal. Hence projections based on these levels could overestimate the impact of reducing personnel and attrition. However, as we just observed, unit conversions (including inactivations) affected only a minority of soldiers in this dataset. The rates of attrition and job turbulence (soldiers who change MOS) shown in Tables 4.1 and 4.2 are in line with historical averages—the overall rate of attrition during FY94 and FY95 was approximately 22 percent, while the percent of soldiers who changed MOS was approximately 13 percent (calculated by dividing the number of soldiers who change MOS during a year by the starting force size).

Table 4.7

Transition Probabilities During FY94

	State at Start of FY95					
			NDMOSQ		DMOSQ	
State at Start of FY94	Not Present	Need IET	Same MOS	Changed MOS	Same MOS	Changed MOS
Not present	0.0	0.3012	0.1552	0.0	0.5437	0.0
Need IET	0.2671	0.1019	0.0166	0.0390	0.5276	0.0478
NDMOSQ	0.2297	0.0116	0.3023	0.1074	0.1749	0.1741
DMOSQ	0.1960	0.0075	0.0026	0.0608	0.6870	0.0461

Table 4.8

Transition Probabilities During FY95

	State at Start of FY96					
			NDMOSQ		DMOSQ	
State at Start of FY95	Not Present	Need IET	Same MOS	Changed MOS	Same MOS	Changed MOS
Not present	0.0	0.2983	0.1934	0.0	0.5083	0.0
Need IET	0.3114	0.1370	0.0181	0.0443	0.4541	0.0351
NDMOSQ	0.2506	0.0065	0.3532	0.1080	0.1528	0.1288
DMOSQ	0.2104	0.0074	0.0045	0.0758	0.6560	0.0459

steady-state probabilities for each of the defined states. These steady-state probabilities estimate the proportion of the force that are in each state in the future.

For this simple Markov model, we use the average of the transition probabilities for FY94 and for FY95. Using these averages, we estimate that the future steady-state DMOSQ rate for the RC will be 74.8 percent of assigned personnel. This projected DMOSQ rate, along with the FY94 to FY96 rates calculated from the SIDPERS data (and originally shown in Table 4.3), is shown in Table 4.9.

Table 4.9

Actual and Projected DMOSQ Rates

Start of Fiscal Year	DMOSQ Rate
1994	75.3%
1995	77.9%
1996	75.3%
Future	74.8%

IMPACT OF REDUCING PERSONNEL MOVEMENTS ON ATTRITION AND TURBULENCE

As we mentioned earlier, if the driving force behind the turbulence and attrition are personnel movements, then we can take actions to mitigate the problems. Using the baseline transition probabilities in the inventory projection model that estimated a future steady-state DMOSQ rate of 74.8 percent, we can model the impact of changing personnel movements in ways that reduce attrition and turbulence and see what impact it has on the projected DMOSQ rate. To do that, we modified the base case transition probabilities and generated new steady-state probabilities for our various states.[12]

Table 4.10 shows the impact of two alternative cases. One case reduces job turbulence, defined as the movement of soldiers qualified to unqualified states, by half; the second case adds to that reduction a 25 percent decrease in the attrition of all soldiers in RC units. For the first case, reducing turbulence by 50 percent, the DMOSQ rate increases from the base case value of 74.8 percent to 78.8 percent. In the second case, the DMOSQ rate is further increased to 80.4 percent.

[12]This was done by averaging the FY94 transition probabilities in Table 4.7 with those for FY95 shown in Table 4.8. The average was then iterated to produce steady-state transition probabilities for a force size of 475,000. These transition probabilities provide the basis for estimating DMOSQ and IET training requirements and how they change under varying assumptions about job change and loss rates.

Table 4.10

Impact of Reducing Turbulence and Attrition on DMOSQ Rates

Fiscal Year	DMOSQ Rate
1994	75.3%
1995	77.9%
1996	75.3%
Future base case	74.8%
50% less turbulence	78.8%
50% less turbulence/ 25% less attrition	80.4%

NOTE: Turbulence is defined as the movement of soldiers from an initial duty MOS to a different one.

The findings replicate the results of an earlier RAND Arroyo Center study.[13] As part of a study examining personnel readiness in the Army RC, Arroyo Center researchers constructed a readiness enhancement model to examine, among other things, the contribution of turbulence and attrition to shortfalls in personnel readiness. Although that model was more sophisticated than the one discussed here, ours yielded consistent results when it examined the prospects of the same two strategies discussed above: reducing job turbulence by 50 percent, and reducing turbulence by 50 percent and attrition by 25 percent. In the first case, the earlier model showed a 9 percent increase in DMOSQ rates; in the second case, the model showed nearly a 15 percent increase.

Another way to think about the effect of such improvements in the DMOSQ rate is in terms of reducing training requirements and the costs to train those soldiers. Again, our findings are consistent with (though somewhat smaller than) those of Orvis et al. (1996). Assuming an RC end strength in the future of 450,000 to 500,000 soldiers, 81,000 soldiers would need reclassification training and 39,000 would need IET. Cutting turbulence of DMOSQ soldiers by 50 percent

[13]See Bruce R. Orvis et al., *Ensuring Personnel Readiness in the Army Reserve Components*, Santa Monica, CA: RAND MR-659-A, 1996.

yields a DMOSQ rate of 78.8 percent, with decreases in the reclassification training requirement of 19,000 soldiers (with potential savings of $93 million in FY94 dollars) and negligible change in IET requirements and costs.[14] Reducing attrition by 25 percent in addition to the 50 percent reduction in DMOSQ turbulence yields an even higher DMOSQ rate of 80.4 percent. Compared to the base case, the reclassification training requirement is reduced by 20,500 soldiers and the number of soldiers needing IET is reduced by 6,500. This leads to potential savings of approximately $190 million (in FY94 dollars).

Understanding the actual effects of such turbulence and attrition reduction strategies requires testing the strategies in a controlled setting that will allow the costs/savings to be systematically evaluated and the uncertainties to be resolved. The RAND research cited above recommended such a strategy. The authors recommended personnel turnover–reduction bonuses for both turnover and attrition strategies ($250 and $900 per eligible soldier, respectively) and proposed a pilot test to evaluate the bonuses.[15] If such programs were implemented, they could provide a very useful means for lowering training demand and helping the TASS achieve further improvements in efficiency and performance.

---

[14]The savings assume the average course costs per student for MOS reclassification training—both direct costs (such as school staff pay and allowances) and indirect costs (such as installation staff pay and allowances)—are about $4,900 in the RC in FY94. For IET, the savings assume an average course cost of approximately $13,900 in FY94 ($6,148 for basic training plus $7,729 for advanced individual training). See Orvis et al. (1996) for more details on how the average course cost per student is derived.

[15]See Orvis et al. (1996).

Chapter Five

CONCLUSIONS AND RECOMMENDATIONS

In this chapter we highlight major issues related to the management of training requirements and delivery of training in the TASS, based on the results presented in previous chapters.

Generally, the results presented in this report indicate that progress is being made in the prototype and within the broader system in managing training requirements and delivering the training needed to meet them. At the same time, the scope of problems that remain warrants continuing vigilance and effort to solve them. These problems are common to both of the areas examined in this report—NCO education and reclassification training conducted in RC schools.

A key problem that lies within the power of the Army school system to solve has to do with the loss of available capacity to meet training requirements. The system is wasting about a third of the training seats that are allotted to deliver training. This is particularly troublesome given the large gap between training requirements and available training seats.

These quotas are being lost for several reasons. One is that unit personnel responsible for making and monitoring reservations for training seats are not yet fully proficient in using the Army's reservation system (ATRRS); hence, they are not making reservations that are needed. Responsible Army agencies (including the U.S. Army's Office of the Deputy Chief of Staff for Personnel, the U.S. Army Reserve Command, and the Army National Guard) are providing additional training and assistance in using ATRRS. These efforts should be maintained.

Quotas are also lost when key resources needed to conduct a course are absent. These resources include qualified instructors and the equipment and facilities, identified in the course's program of instruction, necessary for conducting the course to the Army standard. Organizations within the TASS responsible for coordinating training, such as TRADOC's Regional Coordinating Elements (RCEs), can help ensure that these key resources are located and available. And, as discussed elsewhere,[1] the policies that govern instructor certification and management are critically important for ensuring that qualified instructors in sufficient numbers are available throughout the school system.

Another reason why quotas are lost is simply that some soldiers who make reservations do not show up. It should be expected that some RC soldiers will be unable to keep their reservations because of bad health, family problems, and unanticipated conflicts with civilian employment. Others, however, might attend if additional command emphasis and oversight were directed toward making, keeping, and changing initial reservations and ensuring that available quotas are used to the fullest extent possible. And for those who cannot attend, prompt notification can permit other soldiers on wait-list status to use the vacant seats.

Policies governing quota management are also important in this regard. In our interviews, we often heard that soldiers who might have otherwise attended were "locked out" of courses because not enough school seats were allocated to their unit or, in other cases, were reserved by soldiers who did not show up. One possible way to ensure better quota fill is to consider selective overbooking, as is done in the airline industry to ensure that planes fly at full capacity. Another option includes expanded wait-listing (with better notification when wait-lists become reservations), and earlier reassignment of quotas from units who are not filling them to others that will.

In addition to these problems with quota utilization, our analyses emphasize the importance of reducing training requirements, an area that lies outside traditional school system boundaries. The problem is one of both accuracy and size. A key issue for ensuring

---

[1]See, for example, Winkler et al. (1996).

that schools can provide necessary training is knowing what training is needed. Currently, the Army forecasts its training requirements several years in advance, and while adjustments are made one year prior to execution, this can still lead to shortages and misallocations of training resources to meet the current training requirement. Hence, the Army needs to put policies and programs in place to better forecast training requirements. In our research, we found that it is possible to develop reasonable current estimates of reclassification training requirements using SIDPERS and ATRRS. This tool can also be used to make short-term forecasts, based on historical experience, of the number of soldiers who will need training in the various CMFs and MOSs. Army organizations responsible for managing training requirements have begun to develop and apply such tools.[2]

Finally, efforts should be made to reduce the size of training requirements. The Army's "select-train-promote" policy is one example of how to reduce training requirements, in this case by limiting NCO education to only those NCOs who are selected for promotion. By extension, priorities could be established for determining which soldiers should be sent to reclassification training. Such training might be limited, for example, to soldiers in high-priority units, to those with a minimum remaining service obligation, or to those in selected MOSs. Remaining soldiers might be sent to training as resources are available, or they might be qualified through other means (e.g., structured on-the-job training).

In conclusion, we observe that the Army's problems with managing training requirements and delivering training to meet these requirements are sizable in magnitude and remain a significant challenge for the Army to resolve. The new TASS, while making headway against these problems in some respects, still confronts systemic and structural difficulties. The magnitude of the improvements we observed in Region C and elsewhere are quite modest compared to the scope of the problems observed throughout the entire system. Hence, the solutions to these problems must be broad and sustained in nature, affecting personnel and training policies and practices throughout the Army school system in an integrated fashion. More-

---

[2]New tools employing methodology similar to RAND's are being developed and fielded in the ARNG and USAR.

throughout the Army school system in an integrated fashion. Moreover, continued command emphasis and new policies and programs addressing these problems throughout the TASS will be needed to assure continued progress.

Appendix A

VALIDATING SIDPERS-BASED ESTIMATES OF TRAINING REQUIREMENTS

In FY95, we validated our SIDPERS-based estimates of training requirements with the help of First Army teams, who collected data during Operational Readiness Evaluations (OREs) of 18 units.

APPROACH

The intent was to validate those estimates of soldiers who (a) were not qualified for their DMOS or (b) had not fully completed the NCO education course required for their grade (PLDC, BNCOC, or ANCOC if the soldier was an E-5, E-6, or E-7, respectively). For these validations, the ORE teams examined soldiers' personnel records and, based on information therein, determined if they were DMOSQ and NCOQ (among E-5s through E-7s). Subsequently, we compared these assessments against the information in our SIDPERS files.

RESULTS

The result of this validation is shown in Figure A.1. The left half of the figure shows that among 1,281 soldiers compared with respect to DMOSQ, the ORE and SIDPERS results agreed exactly on the qualification status (DMOSQ or non-DMOSQ) for 1,062 soldiers (83 percent). The right half of the figure shows the same level of agreement for NCO qualification (83 percent, or 442 of 553 soldiers in grades E-5 through E-7).

In those areas where the ORE and SIDPERS results differ, we also observed that many of the differences cancel out. For example, in re-

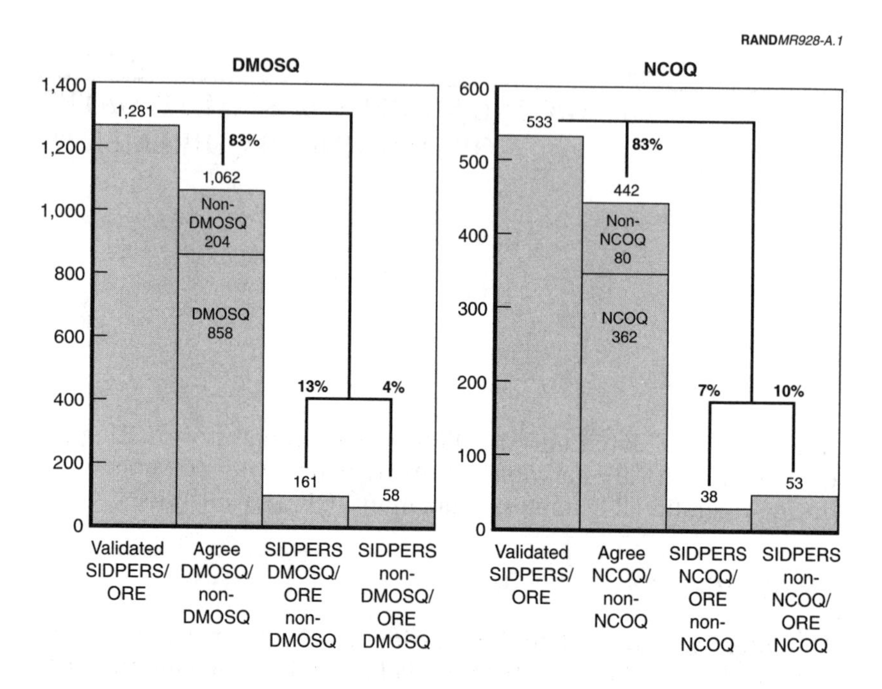

Figure A.1—SIDPERS Accurately Portrays Aggregate Training Requirements

viewing soldiers' DMOSQ, in 161 cases (12.6 percent), SIDPERS found the soldiers DMOSQ but the ORE team did not. In the remaining 58 cases (4.5 percent), SIDPERS found the soldiers to be non-DMOSQ but the ORE team ruled them DMOSQ. Hence, SIDPERS actually *overstates* the rates of duty qualification.

In examining whether NCOs had completed the NCO course required for their grade, as noted above, SIDPERS and the ORE team agreed in 83 percent of cases. When they disagreed, SIDPERS showed 53 cases (9.9 percent) as nonqualified and 38 cases (7.1 percent) as qualified, while the ORE team found the opposite. Hence, SIDPERS slightly *understates* levels of NCO qualification.

The net effect is that SIDPERS estimates of DMOSQ training requirements are slightly lower than ORE estimates, while SIDPERS estimates of NCO training are slightly higher than ORE estimates.

These differences are small, however, indicating that SIDPERS can provide reasonable estimates of potential trainee populations at the aggregate levels (i.e., for total soldiers and potentially within broad groupings like career management fields).

Appendix B

MEASURES OF TRAINING REQUIREMENTS AND SCHOOL DELIVERY

The data to support the analysis described in this report came from a number of separate Army data sets. Data on MOS and NCOES training requirements came from the SIDPERS-USAR and SIDPERS-ARNG systems. Validation of SIDPERS data was done by examining the Military Personnel Records Jacket (MPRJ) records of a sample of soldiers. Data related to RCTI training and IET training in active component training institutions were taken from the ATRRS system. ATRRS provided information on the capacity and production of RCTIs, as well as a record of the training of individual soldiers. Cost-of-travel information for the USAR was taken from the DOLFINS systems. Data related to the status, activations, inactivations, and structure changes of USAR and ARNG units were obtained from the MFORCE files.

The remainder of this appendix contains a more detailed description of each of these data sources and is intended to assist the Army by identifying ways to monitor training requirements and the continued performance of the TASS.

SIDPERS DATA

SIDPERS contains information about individual soldiers in the USAR, ARNG, and the Active Army. (Note that the Army is currently converting SIDPERS for all components to the Total Army Personnel Data Base, TAPDB, system. Future analysis of individual data will have to use TAPDB or, for historical purposes, a combination of TAPDB and SIDPERS.)

SIDPERS-USAR and SIDPERS-ARNG are maintained for the RC by General Research Corporation, Reserve Systems. Data for soldiers in each of the components are maintained in separate data bases. In principle, SIDPERS is a real-time system designed to reflect the current status of Army personnel. In practice, however, there is an indeterminate delay in the time between an actual personnel change and an update to SIDPERS. Thus, a SIDPERS file created at any point in time will be a somewhat "out-of-focus" snapshot of the state of the Army at that point. However, comparisons between two successive snapshots can provide a reasonably accurate way to assess changes occurring over time in the personnel system.

RAND Arroyo Center received data tapes for each RC from General Research Corporation containing SIDPERS data at the end of FY93, FY94, and FY95. For this TASS project, we created a series of combined SIDPERS files that combined data for two fiscal years and both RC. (See the section on computer programs for more details on processing.)

Table B.1 shows the variables for the combined (USAR + ARNG) SIDPERS files. Each file contains a set of variables for each of the two fiscal years represented in the file.

RAND-Created Variables

The combined file contains several variables created at RAND for use on the TASS project. Specifically, they include state variables—which indicate status at a point in time—and change variables—which indicate a change in status between two points in time.

State variables. These include the following:

- TASS region. TASS regions (C, E, and Other) based on the unit state:
  - Region C
    - Florida
    - Georgia
    - North Carolina
    - South Carolina

— Region E
  Illinois
  Indiana
  Michigan
  Minnesota
  Ohio
  Wisconsin

- TASS functional area. Based on a soldier's duty MOS. See Table B.3 at the end of this appendix.
- High-priority unit flag. This flag was created from a list of USAR units obtained at the Office of the Chief, Army Reserve (OCAR) and ARNG units. High-priority units are USAR CFP/FSP units and ARNG Roundout Brigades.
- DMOSQ status. Indicates if a soldier is qualified in his duty MOS. This was produced by comparing DMOS to PMOS, SMOS, and AMOS. A soldier is counted as qualified if his DMOS matched any of these and if his primary MOS skill level was greater than zero (indicating that he had completed IET and been awarded an MOS). If his DMOS does not match any of these, he is counted as unqualified.
- Needs reclassification training. Indicates if a soldier found to be not duty-MOS-qualified needs reclassification training. A soldier is shown as needing reclassification training if his DMOS did *not* match his PMOS, SMOS, or AMOS, and if his primary MOS skill level was greater than zero (indicating that he had previously been awarded an MOS; hence he appears to be a prior-service soldier who could attend a reclassification training course).
- Needs initial entry training. Indicates if a soldier initially found as DMOSQ or non-DMOSQ needs initial entry training. A soldier is shown initially as needing IET if his primary MOS skill level was equal to zero, irrespective of the DMOS and PMOS/SMOS/AMOS match.

Table B.1

Variables for Combined SIDPERS Files

Variable	Description
AMOS	Three-Digit Additional MOS
ASIP	Additional Skill Indicator–PMOS
ASTPGM	ARNG Active Status Program
COMP	Component
DFA+	TASS Functional Area
DIEMS	Date of Initial Entry into Military Service
DIERF	Date of Initial Entry Reserves
DMOS	Three-digit Duty MOS
DMOSAS	Duty MOS ASI
DOR	Date of Rank
DSL	Duty MOS Skill Level
DUTYQ	Duty Qualification Code
HIPRI+	Unit Priority Flag
HSTATE	Home State
MILEDC	Military Education Completed
MILEDE	Military Education Enrolled
MILPC	Military Personnel Class
MOSQ	DMOSQ Status
MUSARC	Assigned MUSARC
NCOQ	NCOQ Status
PAY	Pay Grade
PEBD	Pay Entry Basic Date
PMOS	Three-digit Primary MOS
PMOSAB	PMOS Basis for Acquisition
POSN	Position Number
PSL	Primary MOS Skill Level
RCC	Reserve Category
RECODE	Recode
RECSTA	Record Status
REGION+	TASS Region
SMOS	Three-digit Secondary MOS
SSN	Social Security Number
TAFMS	Total Months Active Federal Service
TECHSV	ARNG Technician Code
TRC	Training/Pay/Retired Category
UIC	Unit Code
USTATE	Unit State
ZIP	Home Zip Code

NOTE: Variables noted with a + were created at RAND.

We found, however, that the information contained in the PSL field was not always consistent with other personnel data and could not be solely relied upon to indicate whether a soldier needs IET. Specifically, it identified some soldiers as needing IET (PSL equal to zero) when other data suggested that they had previously been awarded MOSs.[1] Other soldiers are shown as "needing IET" at the same time that ATRRS showed them having completed IET at an active duty school. For analytic purposes, such soldiers are reassigned into other categories, e.g., as DMOSQ or needing reclassification training, depending on the DMOS and PMOS/SMOS/AMOS match.

Given these findings, we further reviewed the cases in which soldiers were found to be MOS qualified or to need reclassification training in order to confirm that they were prior-service personnel. A handful of these soldiers appear to be new entrants to the RC without prior active duty experience and no record of completing IET at an AC training institution (based on SIDPERS data showing PEBD in the current year and less than three months in TAFMS and no record of IET completion in ATRRS). These soldiers are reassigned to the "need IET" category. Altogether, these changes served to recategorize approximately 22,000 cases (about 4 percent of assigned personnel) from their initial states. The most important change was to reduce the number of soldiers "needing IET" by about 40 percent from the initial estimate based solely on the skill field, which increased the number of soldiers shown as DMOSQ and needing reclassification training to levels shown in the body of the report.

Change variables. These are used in the analyses of DMOSQ and NCOES training. They are not stored in the SIDPERS file; rather, they are calculated by programs that create data tables. (See "Computer Programs" section below.) Change variables include the following:

- Qualification and training status. Determined by comparing soldiers' DMOSQ and training status in successive years, e.g., to indicate if DMOSQ soldiers remain DMOSQ or become

[1]The soldier is determined to have previously received an MOS based on pay grade (e.g., when pay grade is E-5 or greater) or TAFMS and PEBD (i.e., the soldier has been paid for service before the current year and has enough active service time to imply being trained).

NDMOSQ and whether soldiers needing IET or reclassification remain unqualified or become DMOSQ.

- Change in DMOS. Determined by comparing soldiers' DMOS in successive years to determine if it has changed or remained the same.
- Promotion status. Determined by comparing pay grade at the end of two fiscal years.
- NCOES received. Comparison of military education completed (MILEDC) in successive years.

ATRRS DATA

ATRRS is an on-line system used to manage all facets of Army training. One part of ATRRS is a central class reservation system, which is used to reserve seats at training classes and track the progress of students from reservation to graduation. It is this part of the ATRRS system we tapped into in building two separate training files for the TASS project: (1) the class data set and (2) the individual data set.

The Class Data Set

ATRRS generates a number of standard reports that summarize the flow of students through RCTIs. For the current research, we created a school-level data set describing the capacity and throughput of RCTIs by downloading two of these files and combining them at the class level. This file is known as the SUMR6 file, since the ATRRS report of the same name is the starting point for the file. The reports used in the creation of SUMR6 are as follows:

- SUMR6. Classes in RCTIs by quota source with start and end dates, quotas (scheduled seats in a class), reservations, waits, seats available, input, and graduates. SUMR6 provides these data for each class:
  - School code
  - Class name
  - Class phase

— Start and end dates

— Number of spaces allocated

— Number of reserve

— Number on waiting list

— Number of available seats

— Number starting class

— Number of graduates.

- SCHEDULE. Report of selected class schedule data. This report was used to obtain:

  — Class disposition flag (for canceled or not-conducted classes)

  — Course name

  — Number of no shows

  — Number arrived unqualified

  — Class location.

In addition, the "verifrpt" file was used to create several variables on the SUMR6 file. Verifrpt contains lists used to validate the contents of data fields and aggregate variables to higher levels. For example, the school command code variable is derived from a verifprt table matching school code to command.

The final version of ATRRS class file contains the following variables used to determine school capacity and production. These data are presented for each occurrence of a class.

- ALLOC—number of quota allocations. This is the number of quota slots allocated to a particular class. It comes out of the SMDR process and reflects (in theory) the capacity and demand for training.

- RESERV—number of reservations. This is the number of reservations made. Up to 45 days before the start of a class, reservations may be made only if the individual's unit holds a quota allocation for the class. After that time, any person may make a reservation for a class.

- WAIT—number of people on wait-list. A person would be wait-listed for two reasons: (1) there were no remaining allocations and (2) the person's unit did not hold a quota allocation for a course.
- INPUT—number of inputs. This is the number of people who show up for and start the class.
- GRADS—number of graduates. The number of people who successfully complete the class.
- NOSHOWS—number of no-shows of those who had a reservation.
- NOTQUAL—number who arrived not qualified. The variable does not indicate a specific reason.
- CLSFLAG—class status flag. This indicates if a course took place as scheduled.
- CRS—the course name.

From these data, we can derive the following:

- COMMAND—ARNG, USAR, or other. This is obtained from a school code to command table found in verifrpt.
- MOS—the MOS of the course, for MOSQ and NCOES Phase 2.
- CRSCLASS—course classification. The categories are MOSQ, PLDC, BNCOC, ANCOC, with all but PLDC being further categorized into Phases 1 and 2 (and sometimes 3 and higher for MOSQ classes). The last two are derived by parsing the course name.
- CANCELS—taken together, the number of quota allocations, seats, lost because a course did not take place. The distinction between cancelled and nonconducted classes is the time at which the decision was made to drop the class.
- NETQ—net quotas. Number of seats available in classes after dropping cancelled and nonconducted classes.

$$NETQ = ALLOC - CANCELS - NONCNDT$$

- Reservation-to-quotas ratio.

RES_ALL = RESERV / ALLOC

- Ratio of inputs to quotas.

INP_ALL = INPUT / ALLOC

- Number of classes conducted. A count of the number of classes with inputs greater than zero.
- Mean class size. The mean of INPUT over all classes of interest.
- Graduation rate. The ratio of those who show up for and start the class to those who finish:

GRADRATE = GRADS / INPUT

- Walk-ons.

WALKONS = INPUT – RESERV + NOSHOWS + NOTQUAL

- TASS functional area. MOS-based aggregation of courses. The same categories were used for ATRRS and SIDPERS.

These measures were used in both the MOSQ and NCOES analysis. They were calculated by summing each input field to the appropriate level of analysis—functional area for MOSQ courses, and NCO level for NCOES courses—and then calculating the ratios as shown above.

The ATRRS Individual Data Set

This file was created from a tape supplied by the ATRRS contractor (ASM Research) and contains a record of each soldier's actions relative to a course. There is a separate file for each fiscal year. Variables on the files are shown in Table B.2.

Table B.2

Variables for ATRRS Individual Data Set

Variable	Description
YEAR	Fiscal Year
SCHCODE	School Code
CRS	Course Code
PHASE	Course Phase
CLS	Class Number
SSN	Student Social Security Number
QUOTA	Quota Source
COMPNENT	Component of Student
RESSTAT	Reservation Status
INSTAT	Input Status
OUTSTAT	Output Status
REASON	Reason Code
REMARK	Remarks
NAME	Student Name
GENDER	Gender
PG	Pay Grade
PMOSA	Primary MOS

MFORCE DATA

An MFORCE extract obtained from OCAR was used to determine the status of units at the four-digit unit identification code (UIC) level. Two status variables were defined:

- Activations. Indicates if a unit was activated or inactivated within an interval (1 or 2 fiscal years).
- Structure changes. Indicates unit conversions such as branch, size, and series (according to SRC) during an interval.

COMPUTER PROGRAMS

All programming for the TASS project was done using the Statistical Analysis Sytem (SAS).

The TASS project developed a category hierarchy based on three-digit MOS. Table B.3 shows the MOS to TASS functional area conversions.

Table B.3

MOS to TASS Functional Area Conversion

MOS	Functional Area	MOS	Functional Area	MOS	Functional Area
00B	COMBAT SPT	16P	COMBAT ARMS	29M	COMBAT SPT
00E	OTHER	16R	COMBAT ARMS	29N	COMBAT SPT
00R	OTHER	16S	COMBAT ARMS	29S	COMBAT SPT
00Z	OTHER	16T	COMBAT ARMS	29T	COMBAT SPT
01H	HEALTH SERVICE	16Z	COMBAT ARMS	29V	COMBAT SPT
02B	OTHER	17B	COMBAT ARMS	29W	COMBAT SPT
02C	OTHER	18B	OTHER	29X	COMBAT SPT
02D	OTHER	18C	OTHER	29Y	COMBAT SPT
02E	OTHER	18D	OTHER	29Z	COMBAT SPT
02F	OTHER	18E	OTHER	31C	COMBAT SPT
02G	OTHER	18F	OTHER	31D	COMBAT SPT
02H	OTHER	18Z	OTHER	31F	COMBAT SPT
02J	OTHER	19D	COMBAT ARMS	31G	COMBAT SPT
02K	OTHER	19E	COMBAT ARMS	31K	COMBAT SPT
02L	OTHER	19K	COMBAT ARMS	31L	COMBAT SPT
02M	OTHER	19Z	COMBAT ARMS	31M	COMBAT SPT
02N	OTHER	23R	COMBAT ARMS	31N	COMBAT SPT
02T	OTHER	24C	COMBAT ARMS	31P	COMBAT SPT
02U	OTHER	24G	COMBAT ARMS	31Q	COMBAT SPT
02Z	OTHER	24H	COMBAT SVC SPT	31S	COMBAT SPT
11B	COMBAT ARMS	24K	COMBAT SVC SPT	31T	COMBAT SPT
11C	COMBAT ARMS	24M	COMBAT ARMS	31U	COMBAT SPT
11H	COMBAT ARMS	24N	COMBAT ARMS	31V	COMBAT SPT
11M	COMBAT ARMS	24R	COMBAT ARMS	31W	COMBAT SPT
11Z	COMBAT ARMS	24T	COMBAT ARMS	31Y	COMBAT SPT
12B	COMBAT SPT	25L	COMBAT ARMS	31Z	COMBAT SPT
12C	COMBAT SPT	25M	COMBAT SPT	33R	COMBAT SPT
12F	COMBAT SPT	25P	COMBAT SPT	33T	COMBAT SPT
12Z	COMBAT SPT	25Q	COMBAT SPT	33V	COMBAT SPT
13B	COMBAT ARMS	25R	COMBAT SPT	33Y	COMBAT SPT
13C	COMBAT ARMS	25S	COMBAT SPT	34C	COMBAT SPT
13E	COMBAT ARMS	25V	COMBAT SPT	35G	HEALTH SERVICE
13F	COMBAT ARMS	25Z	COMBAT SPT	35H	COMBAT SVC SPT
13M	COMBAT ARMS	27B	COMBAT SVC SPT	35M	COMBAT ARMS
13P	COMBAT ARMS	27E	COMBAT SVC SPT	35P	COMBAT ARMS
13R	COMBAT ARMS	27F	COMBAT SVC SPT	35U	HEALTH SERVICE
13Z	COMBAT ARMS	27G	COMBAT SVC SPT	35Y	COMBAT SVC SPT
14D	COMBAT ARMS	27H	COMBAT SVC SPT	36L	COMBAT SPT
14J	COMBAT ARMS	27J	COMBAT SVC SPT	36M	COMBAT SPT
14R	COMBAT ARMS	27K	COMBAT SVC SPT	37F	OTHER
14S	COMBAT ARMS	27M	COMBAT SVC SPT	38A	OTHER
16D	COMBAT ARMS	27V	COMBAT SVC SPT	39B	COMBAT SVC SPT
16E	COMBAT ARMS	27X	COMBAT SVC SPT	39C	COMBAT SPT
16F	COMBAT ARMS	27Z	COMBAT SVC SPT	39D	COMBAT SPT
16H	UNKNOWN	29E	COMBAT SPT	39E	COMBAT SPT
16J	COMBAT ARMS	29J	COMBAT SPT	39G	COMBAT SPT

Table B.3 (continued)

MOS	Functional Area	MOS	Functional Area	MOS	Functional Area
39L	COMBAT SPT	62E	COMBAT SPT	68G	COMBAT ARMS
39V	COMBAT SPT	62F	COMBAT SPT	68H	COMBAT ARMS
41C	COMBAT SVC SPT	62G	COMBAT SPT	68J	COMBAT ARMS
42C	HEALTH SERVICE	62H	COMBAT SPT	68K	COMBAT ARMS
42D	HEALTH SERVICE	62J	COMBAT SPT	68L	COMBAT ARMS
42E	HEALTH SERVICE	62N	COMBAT SPT	68N	COMBAT ARMS
43E	COMBAT SVC SPT	63B	COMBAT SVC SPT	68P	COMBAT ARMS
43M	COMBAT SVC SPT	63D	COMBAT SVC SPT	68Q	COMBAT ARMS
44B	COMBAT SVC SPT	63E	COMBAT SVC SPT	68R	COMBAT ARMS
44E	COMBAT SVC SPT	63G	COMBAT SVC SPT	68X	COMBAT ARMS
45B	COMBAT SVC SPT	63H	COMBAT SVC SPT	71C	COMBAT SVC SPT
45D	COMBAT SVC SPT	63J	COMBAT SVC SPT	71D	COMBAT SVC SPT
45E	COMBAT SVC SPT	63N	COMBAT SVC SPT	71E	COMBAT SVC SPT
45G	COMBAT SVC SPT	63R	COMBAT SVC SPT	71G	HEALTH SVCS
45K	COMBAT SVC SPT	63S	COMBAT SVC SPT	71L	COMBAT SVC SPT
45L	COMBAT SVC SPT	63T	COMBAT SVC SPT	71M	COMBAT SVC SPT
45N	COMBAT SVC SPT	63W	COMBAT SVC SPT	72E	COMBAT SPT
45T	COMBAT SVC SPT	63Y	COMBAT SVC SPT	72G	COMBAT SPT
45Z	COMBAT SVC SPT	63Z	COMBAT SVC SPT	73C	COMBAT SVC SPT
46Q	OTHER	65B	COMBAT SVC SPT	73D	COMBAT SVC SPT
46R	OTHER	65C	UNKNOWN	73Z	COMBAT SVC SPT
46Z	OTHER	65D	COMBAT SVC SPT	74C	COMBAT SPT
51B	COMBAT SPT	66E	UNKNOWN	74D	COMBAT SPT
51G	COMBAT SPT	66F	UNKNOWN	74F	COMBAT SPT
51H	COMBAT SPT	66G	COMBAT ARMS	74Z	COMBAT SPT
51K	COMBAT SPT	66H	COMBAT ARMS	75B	COMBAT SVC SPT
51M	COMBAT SPT	66J	COMBAT ARMS	75C	COMBAT SVC SPT
51R	COMBAT SPT	66N	COMBAT ARMS	75D	COMBAT SVC SPT
51T	COMBAT SPT	66R	COMBAT ARMS	75E	COMBAT SVC SPT
51Z	COMBAT SPT	66T	COMBAT ARMS	75F	COMBAT SVC SPT
52C	COMBAT SVC SPT	66Y	COMBAT ARMS	75Z	COMBAT SVC SPT
52D	COMBAT SVC SPT	67G	COMBAT ARMS	76C	COMBAT SVC SPT
52E	COMBAT SPT	67H	COMBAT ARMS	76J	HEALTH SERVICE
52F	COMBAT SVC SPT	67J	UNKNOWN	76P	COMBAT SVC SPT
52G	COMBAT SPT	67K	UNKNOWN	76V	COMBAT SVC SPT
52X	COMBAT SVC SPT	67N	COMBAT ARMS	76X	COMBAT SVC SPT
54B	COMBAT SPT	67R	COMBAT ARMS	76Y	COMBAT SVC SPT
55B	COMBAT SVC SPT	67S	COMBAT ARMS	76Z	COMBAT SVC SPT
55D	COMBAT SVC SPT	67T	COMBAT ARMS	77F	COMBAT SVC SPT
55G	COMBAT SVC SPT	67U	COMBAT ARMS	77L	COMBAT SVC SPT
55R	COMBAT SVC SPT	67V	COMBAT ARMS	77W	COMBAT SVC SPT
55X	COMBAT SVC SPT	67Y	COMBAT ARMS	79D	OTHER
55Z	COMBAT SVC SPT	67Z	COMBAT ARMS	81B	COMBAT SPT
57E	COMBAT SVC SPT	68B	COMBAT ARMS	81C	COMBAT SPT
57F	COMBAT SVC SPT	68D	COMBAT ARMS	81E	COMBAT SPT
62B	COMBAT SVC SPT	68F	COMBAT ARMS	81Q	COMBAT SPT

Table B.3 (continued)

MOS	Functional Area	MOS	Functional Area	MOS	Functional Area
81Z	COMBAT SPT	91G	HEALTH SERVICE	93J	COMBAT ARMS
83E	COMBAT SPT	91H	HEALTH SERVICE	93P	COMBAT ARMS
83F	COMBAT SPT	91J	HEALTH SERVICE	93Z	COMBAT ARMS
88H	COMBAT SVC SPT	91K	HEALTH SERVICE	94A	COMBAT ARMS
88K	COMBAT SVC SPT	91L	HEALTH SERVICE	94B	COMBAT SVC SPT
88L	COMBAT SVC SPT	91M	HEALTH SERVICE	94F	HEALTH SERVICE
88M	COMBAT SVC SPT	91N	HEALTH SERVICE	95B	COMBAT SPT
88N	COMBAT SVC SPT	91P	HEALTH SERVICE	95C	COMBAT SPT
88P	COMBAT SVC SPT	91Q	HEALTH SERVICE	95D	COMBAT SPT
88Q	COMBAT SVC SPT	91R	HEALTH SERVICE	96B	COMBAT SPT
88R	COMBAT SVC SPT	91S	HEALTH SERVICE	96D	COMBAT SPT
88S	COMBAT SVC SPT	91T	HEALTH SERVICE	96F	COMBAT SPT
88T	COMBAT SVC SPT	91U	HEALTH SERVICE	96H	COMBAT SPT
88U	COMBAT SVC SPT	91V	HEALTH SERVICE	96R	COMBAT SPT
88V	COMBAT SVC SPT	91X	HEALTH SERVICS	96Z	COMBAT SPT
88W	COMBAT SVC SPT	91Y	HEALTH SERVICE	97B	COMBAT SPT
88X	COMBAT SVC SPT	92A	COMBAT SVC SPT	97E	COMBAT SPT
88Y	COMBAT SVC SPT	92B	HEALTH SERVICE	97G	COMBAT SPT
88Z	COMBAT SVC SPT	92E	HEALTH SERVICE	98C	COMBAT SPT
91A	HEALTH SERVICE	92Y	COMBAT SVC SPT	98D	COMBAT SPT
91B	HEALTH SERVICE	92Z	COMBAT SVC SPT	98G	COMBAT SPT
91C	HEALTH SERVICE	93B	COMBAT ARMS	98H	COMBAT SPT
91D	HEALTH SERVICE	93C	COMBAT ARMS	98J	COMBAT SPT
91E	HEALTH SERVICE	93D	COMBAT ARMS	98K	COMBAT SPT
91F	HEALTH SERVICE	93F	COMBAT ARMS	98Z	COMBAT SPT

REFERENCES

Buddin, Richard J., and David W. Grissmer, *Skill Qualification and Turbulence in the Army National Guard and Army Reserve,* Santa Monica, CA: RAND MR-289-RA, 1994.

Department of the Army Inspector General (DAIG), *Special Assessment of Reserve Component Training,* Washington, D.C., January 11, 1993.

Orvis, Bruce R., Herbert J. Shukiar, Laurie L. McDonald, Michael G. Mattock, M. Rebecca Kilburn, and Michael G. Shanley, *Ensuring Personnel Readiness in the Army Reserve Components,* Santa Monica, CA: RAND MR-659-A, 1996.

Shanley, Michael G., John D. Winkler, and Paul S. Steinberg, *Resources, Costs, and Efficiency of Training in the Total Army School System,* Santa Monica, CA: RAND, MR-844-A, 1997.

Sortor, Ronald E., Thomas F. Lippiatt, J. Michael Polich, and James C. Crowley, *Training Readiness in the Army Reserve Components,* Santa Monica, CA: RAND, MR-474-A, 1994.

Winkler, John D., Michael G. Shanley, James C. Crowley, Rodger A. Madison, Diane Green, J. Michael Polich, Paul Steinberg, and Laurie McDonald, *Assessing the Performance of the Army Reserve Components School System,* Santa Monica, CA: RAND, MR-590-A, 1996.